Michael Nolting

Dynamic Reconfiguration Methods for Active Camera Networks

Michael Nolting

Dynamic Reconfiguration Methods for Active Camera Networks

Coordination Algorithms for Distributed Smart Cameras Networks

Südwestdeutscher Verlag für Hochschulschriften

Impressum/Imprint (nur für Deutschland/only for Germany)
Bibliografische Information der Deutschen Nationalbibliothek: Die Deutsche Nationalbibliothek verzeichnet diese Publikation in der Deutschen Nationalbibliografie; detaillierte bibliografische Daten sind im Internet über http://dnb.d-nb.de abrufbar.
Alle in diesem Buch genannten Marken und Produktnamen unterliegen warenzeichen-, marken- oder patentrechtlichem Schutz bzw. sind Warenzeichen oder eingetragene Warenzeichen der jeweiligen Inhaber. Die Wiedergabe von Marken, Produktnamen, Gebrauchsnamen, Handelsnamen, Warenbezeichnungen u.s.w. in diesem Werk berechtigt auch ohne besondere Kennzeichnung nicht zu der Annahme, dass solche Namen im Sinne der Warenzeichen- und Markenschutzgesetzgebung als frei zu betrachten wären und daher von jedermann benutzt werden dürften.

Coverbild: www.ingimage.com

Verlag: Südwestdeutscher Verlag für Hochschulschriften GmbH & Co. KG
Heinrich-Böcking-Str. 6-8, 66121 Saarbrücken, Deutschland
Telefon +49 681 37 20 271-1, Telefax +49 681 37 20 271-0
Email: info@svh-verlag.de

Approved by: Hannover, Leibniz Universität, Diss., 2011

Herstellung in Deutschland (siehe letzte Seite)
ISBN: 978-3-8381-3218-1

Imprint (only for USA, GB)
Bibliographic information published by the Deutsche Nationalbibliothek: The Deutsche Nationalbibliothek lists this publication in the Deutsche Nationalbibliografie; detailed bibliographic data are available in the Internet at http://dnb.d-nb.de.
Any brand names and product names mentioned in this book are subject to trademark, brand or patent protection and are trademarks or registered trademarks of their respective holders. The use of brand names, product names, common names, trade names, product descriptions etc. even without a particular marking in this works is in no way to be construed to mean that such names may be regarded as unrestricted in respect of trademark and brand protection legislation and could thus be used by anyone.

Cover image: www.ingimage.com

Publisher: Südwestdeutscher Verlag für Hochschulschriften GmbH & Co. KG
Heinrich-Böcking-Str. 6-8, 66121 Saarbrücken, Germany
Phone +49 681 37 20 271-1, Fax +49 681 37 20 271-0
Email: info@svh-verlag.de

Printed in the U.S.A.
Printed in the U.K. by (see last page)
ISBN: 978-3-8381-3218-1

Copyright © 2012 by the author and Südwestdeutscher Verlag für Hochschulschriften GmbH & Co. KG and licensors
All rights reserved. Saarbrücken 2012

Contents

List of Figures   5

List of Algorithms   11

1 Introduction   13
    1.1 Motivation: Active Camera Networks 13
    1.2 Problem Statement and Contribution 16
    1.3 Classification and Scientific Focus 19
    1.4 Overview of the Thesis 21

2 Active Camera Networks   23
    2.1 Definition: Active Camera 23
    2.2 Detecting Targets 26
    2.3 Position Control and Image Acquisition 28
        2.3.1 Sensor Control 29
        2.3.2 Examples for Mobile Entities 30
    2.4 Image Interpretation 33
        2.4.1 Smart Camera Prototypes 34
        2.4.2 Computer Vision 39
    2.5 Summary 42

3 System Model   45
    3.1 Active Cameras 45
        3.1.1 Field of View 47
        3.1.2 Camera's State 49
        3.1.3 Clock Synchronization 49

3.2	Perceiver Nodes		51
	3.2.1	Target Requests	51
	3.2.2	Modeling Perceiver-Observation Uncertainty	52
3.3	Summary		53

4 System Architecture for Active Cameras — 55

4.1	Requirements of Active Camera Networks		56
4.2	Adaptive Location Management Architecture		57
	4.2.1	Architecture Overview	57
	4.2.2	Layer 0: Active Sensing	60
	4.2.3	Layer 1: Communication	61
	4.2.4	Layer 2: Positioning	62
	4.2.5	Layer 3: Coordination	65
	4.2.6	Cross-Layer Event Handler	67
4.3	Summary		68

5 Dynamic Reconfiguration Methods — 69

5.1	Problem Statement: Wide-Area Target Acquisition		70
	5.1.1	Formal Description	71
	5.1.2	Proof of Problem Complexity	74
5.2	DRofACN		76
	5.2.1	Asynchronous Scheduling Process	77
	5.2.2	IDLE Mode	83
	5.2.3	MOVING Mode	83
	5.2.4	OBSERVATION Mode	84
	5.2.5	Correctness	85
	5.2.6	Phenomena Adaptivity (ENRA)	87
5.3	Active Frame Synchronization		90
	5.3.1	Problem Statement: Frame Synchronization	92
	5.3.2	ACFSync: Active Camera Frame Synchronization	92
	5.3.3	Beacon-assisted Clock Synchronization Algorithm	95
	5.3.4	Cooperative Frame Synchronization Algorithm	104
5.4	Summary		111

6 Evaluation   113
6.1 Performance Metrics 113
6.2 DRofACN 118
6.2.1 Experimental Setup 118
6.2.2 Scalability 120
6.2.3 Packet Loss 126
6.2.4 Motion of Targets 128
6.2.5 Target Speed 130
6.2.6 Phenomena Adaptivity (ENRA) 134
6.3 ACFSync (operation mode 1) 136
6.3.1 Experimental Setup 136
6.3.2 Synchronization Accuracy 137
6.3.3 Error Rate 139
6.3.4 Time Complexity 141
6.3.5 CPU and Memory Utilization 144
6.4 ACFSync (operation mode 2) 144
6.4.1 Experimental Setup 144
6.4.2 Noise 145
6.4.3 Perspective 146
6.4.4 Number of Targets 151
6.4.5 Real-world Experiment 151
6.4.6 CPU and Memory Utilization 154
6.5 Summary 155
6.5.1 DRofACN 156
6.5.2 ACFSync 157

7 Related Work   159
7.1 Dynamic Reconfiguration 159
7.1.1 Scheduling 160
7.1.2 Dynamic Vehicle Routing Problem with Time Windows ... 161
7.1.3 Sensor Planning for Visual Surveillance 163
7.2 Operating System and Middleware 165
7.2.1 General-Purpose Middleware 166

	7.2.2	Middleware for Embedded Systems	167
	7.2.3	Middleware for Organic Systems	168
7.3	Active Cameras and Active Vision	169
	7.3.1	Optimal Placement	169
	7.3.2	Active Cameras	171
	7.3.3	Active Vision Agents	173
7.4	Time Synchronization in Sensor Networks	174
	7.4.1	Sender-to-Receiver Synchronization	176
	7.4.2	Receiver-to-Receiver Synchronization	178
7.5	Summary	180

8 Conclusion 183
8.1 Summary of Contributions 184
8.2 Future Research Opportunities 187

List of Figures

1.1 Screenshot of a simulated Active Camera Network at Hannover Main Station. Here, 9 Active Cameras (big triangles) are positioned at the front yard in order to detect targets of interest (small triangles). The number of cameras may vary - for a complete surveillance of the Main Station hundreds of them are needed. Source: Google Earth 14

2.1 Active Camera concept 25
2.2 Measurement variables for a proximity sensor [1] 27
2.3 Caroline - Autonomous car from the University of Braunschweig [2] 31
2.4 AR drone from Parrot A.S. - controllable through Wi-Fi . . . 32
2.5 Bluefin-12 UUV with a Buried Object Scanning Sonar (BOSS) integrated in two wings 33
2.6 Generic architecture of a Smart Camera [3] 35
2.7 Image, camera, and world coordinate frames [4] 39

3.1 System model: Active Cameras, perceiver nodes, and dynamic targets 47
3.2 Geometry of an Active Camera's field of view [5] 48

4.1 Single Active Camera: Adaptive Location Management Architecture 58
4.2 Optimal target-to-camera distance 64

4.3	An example of the location-dependent quality for observing a target in the camera's actuation range	65
4.4	An example of the location-dependent quality for observing a target in the camera's actuation range on the basis of the target's current position	66
4.5	Target Information Message (TIM) and corresponding information of a target. The target's ID is unique and defined by the perceiver node detecting the target's first occurrence.	67
5.1	Two targets of interest (ToIs) are observed (observation condition is true) in an Active Camera Network of two Active Cameras (ACs) in $[t', t'')$.	71
5.2	A Hamiltonian path (red) over a graph.	74
5.3	State machine of the *DRofACN* method	76
5.4	Quality function of target-to-camera distance [6] (red points correspond to computer vision success rates for the x-values respectively and are supporting points for the construction of the quality function)	81
5.5	Quality function of the view angle [7] (red points correspond to computer vision success rates for the x-values respectively and are supporting points for the construction of the quality function)	82
5.6	Computation of the best center of movement	90
5.7	Ratio of client and server single trip times: (a) Asymmetric GPRS network link (b) Ideal symmetric link (line through origin).	91
5.8	Overview about the *ACFSync* method	93
5.9	Conceptual overview about the sender-to-receiver approach based on beacon-assisted clock synchronization	95
5.10	The effect of exposure time on signal sampling	97
5.11	The entire sampling process illustrated	100
5.12	Concept of the beacon-assisted clock synchronization algorithm based on the receiver and sender component	103
5.13	Temporal analysis and masking	105

LIST OF FIGURES

5.14 Example for calculating the frame offset by correlation 106
5.15 Spatial offsets of a person entering the camera's field of view . 107
5.16 Movement detection of an Active Camera 109

6.1 Experiments with targets entering the surveillance area on straight-lines (left picture) or parabolic trajectories (right picture) . . . 118
6.2 ACs are positioned in a grid on the surveillance area. The actuation ranges of neighboring ACs overlap by a quarter of the actuation radius and decreases with the number of ACs. . 119
6.3 Relation of target generation rate and resources needed for achieving a specific TAR ratio: For target acquisition ratios up to 80 %, the number of ACs needed increases proportionally to the number of ToIs. For a TAR > 90 % significantly more resources are needed. 121
6.4 Relation of target generation rate and resources needed for achieving a specific mean target detection time: For a mean target detection time below 20 s the number of ACs increases proportionally to the number of ToIs. For mean detection times below 10 s significantly more ACs are needed. 122
6.5 System size and influence on load 123
6.6 System size and influence on imaging quality 123
6.7 Number of unobserved targets for low target generation rates for an ACN consisting of 16 cameras in relation to the target's distance at time of becoming salient before leaving 124
6.8 Number of unobserved targets for high target generation rates (overload) for an ACN consisting of 16 cameras in relation to the target's distance at time of becoming salient before leaving 125
6.9 AC-to-AC packet loss and the influence on TAR. In case of 100 % of packet loss, the observation condition is not met any more but the TAR ratio only decreases to 80 % of the original value (in scenarios with more than 16 ACs). 127
6.10 Perceiver-to-AC packet loss and influence on TAR. In case of 100 % of packet loss, no target requests reach the ACs any more. 128

6.11 Perceiver-target localization error and its influence on the system's performance: An error of up to $20\,cm$ does not influence the TAR ratio significantly. 129
6.12 Perceiver-target localization error and its influence on the system's performance: An error of up to $20\,cm$ does not influence the mean target detection time significantly. 129
6.13 Evaluation of TAR in relation to the target speed and system size (sensing range: two times the actuation range). 131
6.14 Evaluation of TAR in relation to the target speed and system size (sensing range: five times the actuation range) 132
6.15 Evaluation of the target detection time in relation to the target speed and system size (sensing range: five times the actuation range) 133
6.16 System performance (TAR) without network reconfiguration (target generation rate: $6\,ToIs/s$; the spread represents the trajectory's width and targets enter the surveillance area uniformly distributed over this width) 134
6.17 System performance (TAR) with network reconfiguration (target generation rate: $6\,ToIs/s$; the spread represents the trajectory's width and targets enter the surveillance area uniformly distributed over this width) 135
6.18 Photo of the experimental beacon sending a time stamp every 2 seconds [8] 136
6.19 Zoom levels of the PTZ camera utilized to change the signal area 137
6.20 Experimental setup of a beacon and a Smart Camera capturing the beacon signal. A time stamp is sent by the beacon every 2 seconds. 138
6.21 Histogram of offsets between received time stamp and wall clock 138
6.22 Error rate over signal area (represented by different zoom levels) 140
6.23 Error rate over LED intensity with fixed mask 141
6.24 Normal and over-exposed mode (for zoom level 6x) 142
6.25 Packet reception duration over LED intensity, normal exposure 142
6.26 Packet reception duration over LED intensity, over-exposed . . 143

LIST OF FIGURES

6.27 Person moving with 1.5 $\frac{m}{s}$ from right to left through a camera's field of view 144

6.28 Different noise levels 145

6.29 Frame offset for autocorrelation under the influence of additive normally distributed noise 146

6.30 36 cameras positioned each 10 degrees around the surveillance area 147

6.31 Calculated frame offset between the 0° camera and cameras 10° to 350° (on a circle around the surveillance area) 148

6.32 15 cameras positioned on an arc over the surveillance area vertically to the direction of movement 148

6.33 Calculated frame offset between the 0° camera and 15 cameras positioned on an arc vertically to the direction of movement . 149

6.34 15 cameras positioned on an arc over the surveillance area horizontally to the direction of movement 150

6.35 Calculated frame offset between the 0° camera and 15 cameras positioned on an arc horizontally to the direction of movement 150

6.36 Calculated frame offset between the 0° camera and cameras 10° to 350° on a circle around the surveillance area with two persons (pair scene) 152

6.37 Calculated frame offset between the 0° camera and cameras 10° to 350° on a circle around the surveillance area with 7 persons (crowded scene) 152

6.38 Relation of event duration and frequency of occurrence in an office hallway scenario 154

6.39 Average synchronization error with standard deviation in an office hallway scenario 155

7.1 General-purpose middleware layers [9] 166

7.2 Critical path of traditional time synchronization protocols . . 175

7.3 Overview of related work 182

List of Algorithms

1	Heartbeat Algorithm	62
2	DRofACN - Target Cache Thread	78
3	DRofACN - Asynchronous Scheduling Process	79
4	IDLE Mode	83
5	MOVING Mode	83
6	OBSERVATION Mode	84
7	ENRA - Reconfiguration Process	89
8	Beacon Algorithm (operation mode 1)	102
9	Beacon-AC Algorithm (operation mode 1)	104
10	Local Algorithm of AC_j (operation mode 2)	109
11	AC-AC Algorithm (operation mode 2)	110

Chapter 1

Introduction

"I move, therefore I see." (Hamada 1992)

1.1 Motivation: Active Camera Networks

Organic Computing is based on the insight that we will soon be surrounded by large collections of autonomous systems which are equipped with sensors and actuators [10]. In this respect, sensor systems provide a key feedback to allow autonomous systems to be aware of their environment and react to it. Without sensor-based feedback, these systems could only operate in the most controlled conditions since they could not perceive and respond to changes in their workspace. In order to achieve these goals, these systems would have to act more independently, flexibly, and autonomously. They will have to exhibit life-like properties, i.e. self-x properties like self-organization, self-adaptation, or self-healing. Those systems are called "organic". A good example of using sensors for increasing reliability and flexibility is in Smart Cameras [11, 12]. In contrast to traditional cameras, Smart Cameras are equipped with on-board processing and communication units. The combination of recent advances that have been achieved in the areas of computer vision, image understanding (e.g. object recognition, object tracking, and scene analysis), and computer

14 CHAPTER 1. INTRODUCTION

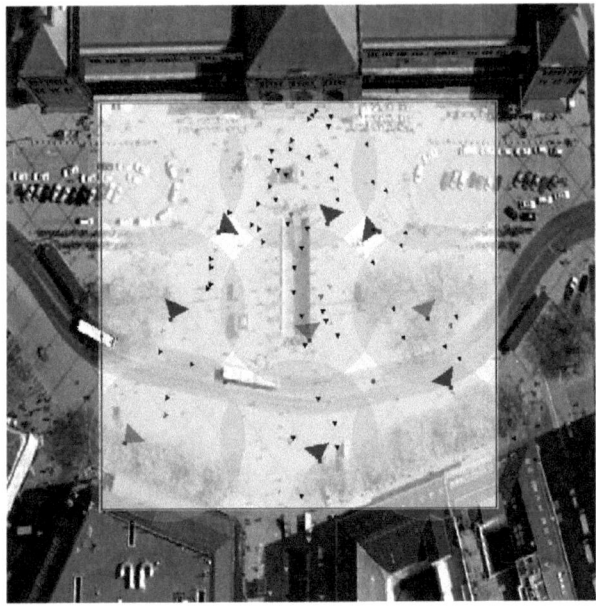

Figure 1.1: Screenshot of a simulated Active Camera Network at Hannover Main Station. Here, 9 Active Cameras (big triangles) are positioned at the front yard in order to detect targets of interest (small triangles). The number of cameras may vary - for a complete surveillance of the Main Station hundreds of them are needed. Source: Google Earth

architecture provide the basis to allow Smart Cameras to detect variations in their workspace by utilizing real-time sensor feedback. Thus, they are able to process incoming vision data on-board in terms of anomalous situations and communicate this to the system's operator.

Currently, camera networks try to cover the entire area or the most important parts of it with a set of passive image sensors. Consequently, they have difficulties in acquiring high-resolution shots selectively. Therefore, system designers have to select the number of cameras and their placement based on *a priori* information considering the requirements of the underlying surveillance

1.1. MOTIVATION: ACTIVE CAMERA NETWORKS 15

task, e.g. number and frequency of targets occurring and the so-called *hot spots* of occurrence. This approach is applicable in controlled and static environments. Nevertheless, *a priori* information becomes less useful in dynamic environments, since dynamics such as a varying number of targets may occur at runtime. The broad range of requirements, that algorithms have for the interpretation of scenes from multiple perspectives, adds up to these difficulties and again increases the number of necessary cameras. One viable and cost-effective alternative to just increasing the number of cameras to the demands of surveillance applications is to make efficient use of position drives, i.e. by adding pan/tilt/zoom drives or attaching cameras to mobile entities such as ground or air vehicles.

Recent advances in the area of robotics have led to the development of autonomous vehicles and unmanned aerial vehicles that can be used to explore operational environments such as urban areas or unknown building structures. For that purpose, such vehicles are equipped with a large number of sensors and actuators. While each autonomous vehicle is able to explore the environment individually, wireless communication between multiple vehicles allows for networking them into a collaborative multi-vehicle system. This makes way for accomplishing goals that cannot be achieved by a single vehicle. As an example consider that several distant parts of the operational environment have to be monitored by sensors at the same time due to several distinct events. Additionally, visual sensors (cameras) and advances in image processing in particular are an important driving force for many applications. In that context, we consider systems of multiple autonomous vehicles - each one equipped with a visual sensor - as *Active Camera Networks*. Applications of such systems are manifold and include, for example, the exploration and surveillance of large areas. Key components to robustly implement such applications are distributed control algorithms that adapt the system's behavior to changing environmental conditions and efficiently coordinate the usage of system resources, in particular the available cameras. In general, we assume that the operational environment cannot be captured by the sensors in the system at once due to its size. However, the mobility of vehicles in the system can be used to dynamically focus the cameras' sensing ranges to different locations over time. One advantage of using a dynamically self-configurable network is that the set-up cost can be reduced significantly. It would be prohibitively expensive to have a static set-up that handles all possible situations. For example, suppose the imaginary network is deployed at Hannover Main Station's front yard as depicted in Figure 1.1 and we need to focus on one person's face in order to perform a biometric task. The person walks around the front yard and the network's goal is to obtain a high-resolution image of this person while also observing other activities going on at the front yard. In order to fulfill this task, an Active Camera Network can dynamically reconfigure the parameters of cameras which are in range of the person. Thus, it is possible to capture high-resolution imagery of the person irrespective of where it is in the front yard. Using a static camera network for this task would be very expensive and a huge waste of resources, both technically and economically.

Smart Cameras extended by activity control are called *Active Smart Cameras*. In the remainder of this thesis, we will simply use the term *Active Camera*, *camera* or *AC* as an abbreviation for Active Smart Camera. The use of camera networks often raises important privacy concerns, e.g. camera-based applications can potentially violate the privacy of observed individuals. Therefore, useful mechanisms for addressing these concerns have to be provided. Nevertheless, this problem has not been addressed in this thesis, but it is clear that it is a source of concern for many people, as information about their private life can be accessed through the network. Due to this reason, this problem is addressed in other research projects (e.g. [13, 14]) explicitly focusing on these socio-ethical issues.

1.2 Problem Statement and Contribution

The main goal of a reconfiguration method is to select the suitable operation modes of the system tasks in order to optimize a certain global objective function. For example, in case of visual surveillance, a global objective function can be defined measuring the successful completion of a face recognition task (e.g. finding subjects from a system-wide watch list). For this purpose, images

1.2. PROBLEM STATEMENT AND CONTRIBUTION 17

have to be captured across the network. In addition, temporal correctness has to be guaranteed to allow for data fusion mechanisms between cameras. Thus, a given quality of service is associated with each task, e.g. there is a difference for system-wide face recognition if a high-quality frontal view image with a correct time stamp is available in contrast to images without any time information showing the back of a person. These different operation modes of an Active Camera capturing an image of a moving target can be expressed by multiple configurations, each configuration exhibiting a system benefit and requiring a different reconfiguration cost, e.g. the cost for repositioning the Active Camera.

Nevertheless, visual surveillance takes place in dynamic environments and must provide mechanisms to detect and identify motion image patterns of moving targets, e.g. objects like humans. This is achieved by sophisticated computer vision algorithms making way for target detection, object association, and data fusion across multiple cameras. The monitored targets may activate different variants for image processing, each one providing a given quality of service and system benefit. The target type, its distance to the camera, speed, or the current luminance are examples of parameters of a dynamic environment that may trigger some appropriate mechanisms. Fault-tolerance requirements are another aspect present in Active Camera Networks. Failure detection or determining the loss of clock synchronization may trigger recovery tasks that must be executed by a given deadline. Additionally, Active Camera Networks may work under eventual overload conditions (e.g. target tracking of hundreds of people) and must be highly adaptive, ensuring temporal correctness while exhibiting graceful degradation.

Perhaps one of the first initiatives to investigate how technical systems can be equipped with so-called *life-like properties* in order to pave the way for adaptability and self-configuration within these systems was the research initiative Organic Computing (DFG SPP 1183) [15]. The Organic Computing initiative aims at overcoming drawbacks of current top-down engineering approaches. Instead of designing a system as a static and thoroughly planned automaton with predefined states and behavior, more flexible approaches are investigated. An Organic Computing system is able to develop and adjust itself to changing environmental influences by adding organic features allowing for:

- self-organization,
- self-adaptation, and
- self-configuration.

These self-x properties can be translated into concrete design features for Active Camera Networks to provide a basis for dynamic reconfiguration. In addition, the integration of these self-x properties into a holistic system architecture is an important aspect of Organic Computing. They have been implemented as part of this thesis as follows:

Active Cameras are able to collaborate on target acquisition in wide-area scenarios autonomously. Therefore, they can be used for capturing images of dynamic targets by reconfiguring their location in terms of their position and orientation. These Active Cameras *self-organize* their location. The user or system administrator can set up constraints (e.g. size and position of actuation ranges), but does not need to supervise this reconfiguration process in detail.

Active Cameras are further able to manage their position so that optimal target acquisition is achieved in wide-area scenarios. For this purpose, they have to collaborate and decide in real-time which target to observe next. In Section 5.1.2, it is shown that the process of acquiring exactly one image of each target entering a surveillance area is an NP-complete derivative of the Hamiltonian Path problem. A distributed control algorithm based on dynamic reconfiguration that helps to find close to optimal solutions to this problem is presented in Section 5.2. This heuristic is an example for an algorithm enabling *self-adaptation* properties for Active Camera Networks.

Self-Configuration in Active Camera Networks implies that Active Cameras are able to detect configuration failures in advance. For example, a potential loss of clock synchronization has to trigger recovery tasks, since synchronized clocks are the basis for data fusion and aggregation across multiple cameras. Section 5.3 introduces a frame synchronization method which is able to maintain Active Cameras' clocks synchronized. Since this method uses visual data for clock synchronization only, it can be used in distributed scenarios where no infrastructure for time synchronization is available.

This thesis introduces a class of dynamic reconfiguration methods that enable Active Cameras to collaborate in surveillance scenarios by relying on Organic Computing features as introduced above. The following section summarizes the major contributions of this work and explains its scientific focus.

1.3 Classification and Scientific Focus

Today's surveillance networks rely on passive cameras. Therefore, system designers have to select the number of cameras and their placement based on *a priori* information considering the requirements of the underlying surveillance task, e.g. number and frequency of targets occurring and so-called *hot spots* of occurrence. The current research focuses on the usage of Active Cameras in order to overcome these placement constraints. This thesis presents a novel system architecture for Active Camera Networks, which is tailored to suit the needs arising in dynamic surveillance scenarios. Active Smart Cameras form a wireless mobile ad-hoc network (MANet) and self-organize their position and orientation. This novel approach to the architecture of Active Camera Networks serves as a basis for dynamic reconfiguration methods which are needed to cope with constantly changing environments and observe a surveillance area collaboratively. The following three main aspects are addressed in this thesis and contribute to dynamic reconfiguration in Active Camera Networks:

1. **Software architecture for Active Camera Networks:** Active Camera Networks as introduced above consist of a high number of Active Cameras in order to cooperatively solve surveillance tasks, which could not be achieved by a single camera or only through considerably more stationary cameras. Our system architecture as presented in Chapter 4 modularizes the various functions needed. Relating mobility, cameras can change their orientation as well as their position. A realistic sensing performance metric is integrated that models the actual coverage characteristics of the camera from a computer vision's perspective.

Thus, the requirements of the underlying computer vision algorithms can be encapsulated into the sensing constraints. The system architecture consists of four layers and runs on each camera independently. This distributed design allows for scalability concerning the number of cameras and adaptability in terms of the environment. These are important properties of an organic system.

2. **Wide-area target acquisition:** A reconfiguration method was developed allowing for dynamic and distributed control of nodes in Active Camera Networks with the goal of capturing high quality images of moving targets. It addresses application scenarios where events unfold over a large geographic area and close-up views have to be acquired for biometric tasks such as face detection. There is no central unit accumulating and analyzing all the data. The overall goal is to capture all targets of interest in the region of deployment of the cameras exactly once, while maintaining a high imaging quality according to the requirements of the underlying computer vision algorithm. Utilizing Active Cameras in such a scenario makes way for efficient use of resources. Nevertheless, this control cannot be based on separate analysis of the sensed imagery in each camera. They must act collaboratively to be able to acquire exactly one capture of each target of interest. Simulations with up to 100 Active Cameras show the scalability and reliability of the proposed method. The performance of different target generation rates is analyzed and it is shown that an Active Camera Network of 100 nodes can handle up to 2,500 targets of interest simultaneously with a target acquisition ratio of 90 % and a mean target detection time of less than 10 seconds.[1] After having shown in Section 5.1.2, that the wide-area target acquisition problem is NP-complete, a heuristic is presented in Section 5.2 to approximate solutions to this problem in real-time.

3. **Active frame synchronization:** In this thesis, novel algorithms for frame-level and visual cue-based clock synchronization have been devel-

---

[1] A network of 100 pan/tilt/zoom cameras, i.e. they are able to pan in order to change their viewing direction but not to move, achieves only a target acquisition ratio of 50 % in such a setting.

oped. This makes way for Active Camera Networks, which are based on distributed vision networks without a centralized server to synchronize the cameras. Two methods are proposed in Section 5.3: (1) a beacon-based clock synchronization method and (2) a cooperative synchronization method where Active Cameras sharing the same field of view synchronize their clocks on the basis of correlating detected salient events. Since both methods utilize optical events, they do not rely on specific hardware other than the visual sensor itself. The beacon-based approach achieves a synchronization accuracy of one frame in 70 % of the cases. Thereby, the accuracy of the method only relies on the visual sensor's sampling rate. If participating cameras capture their environment with a frame rate of 25 frames per second, the beacon-based approach achieves an accuracy of $40\,ms$. A synchronization accuracy within tens of milliseconds is sufficient for scenarios, where visual events are triggered by human beings moving with a velocity in the order of few meters per second. The cooperative approach achieves good synchronization results if the difference of the view angles of both cameras is less than $45°$ or counterpart.

1.4 Overview of the Thesis

Initially, this thesis discusses general aspects of Active Camera Networks before presenting a unified system model for those. Based on this, we present a software architecture for Active Camera Networks allowing for scalability and self-organization. After this, the dynamic reconfiguration methods are presented and evaluated which have been embedded in the distributed software architecture. More precisely, this thesis is structured as follows:

Chapter 2 gives a definition for Active Cameras and a broad overview of their components. We present how activity can support the image acquisition step. For this purpose, examples for mobile entities such as unmanned vehicles are given.

Chapter 3 presents a unified system model for the discussion of aspects related to Active Camera Networks. First, we present how activity can be integrated into camera networks (e.g. by defining the cameras' actuation ranges). We introduce the concept of perceiver nodes (low-cost sensors) interacting with Active Camera Networks for target surveillance. For this purpose, the so-called *target requests* are generated delivering spatio-temporal information of moving targets within the workspace.

In Chapter 4, we show how the requirements of Active Camera Networks can be encapsulated into a decentralized software architecture paving the way for dynamic reconfiguration methods. A four-layered software architecture that is based on Organic Computing's generic *Observer/Controller* concept is presented.

Chapter 5 is devoted to in-detail examinations of the dynamic reconfiguration methods. After formulating the general problem statement, we present a reconfiguration method for wide-area target acquisition and active frame synchronization in detail.

Chapter 6 evaluates the performance of our system architecture and reconfiguration methods. Several performance metrics are introduced and research questions are raised for each algorithm. The evaluation indicates that an Active Camera Network of 100 nodes can handle up to 2,500 targets of interest simultaneously with a target acquisition ratio of 90 % and a mean target detection time of less than 10 seconds. Our time synchronization algorithm achieves a synchronization accuracy of one frame in 70 % of the cases.

Chapter 7 presents related work in the area of dynamic reconfiguration. First, we give a general literature overview and then we review related work for each of our contributions.

Chapter 8 concludes this thesis by summing up the contributions. An outlook on future work is provided.

Chapter 2

Active Camera Networks

Camera networks are undergoing a transition from pure static rectilinear cameras to hybrid solutions that include other sensor types, different camera resolutions, and Active Cameras. In the recent past, pan/tilt/zoom (PTZ) camera networks have been considered for their ability to cover large areas and capture high-resolution information of regions of interest in dynamic scenes. This chapter gives a broad overview about Active Camera Networks. Section 2.1 defines Active Cameras and explains how the control of the camera's position can alleviate the image interpretation. In Section 2.2, we explain from a technical perspective how sensors can detect targets in a surveillance area. In Section 2.3, we present the mechanical hardware which can be used to control a camera's position. In Section 2.4, the electronic hard- and software, i.e. Smart Cameras and computer vision algorithms, are presented which process data captured by a camera. Afterwards, we close with a summary.

2.1 Definition: Active Camera

Active Camera Networks possess mechanisms that are able to actively control camera parameters. Thereby, cameras of such networks are reconfigurable regarding their position and orientation, and indirectly, this includes the baseline (in a two camera system) according to the requirements of the system goal. They may also be equipped with spatially variant (foveal) sensors. More broadly, Active Camera Networks encompass attention, i.e. selective sensing in space, resolution, and time. This is achieved by modifying physical camera parameters, i.e. the way data is captured [16].

Visual data captured by cameras contains information about patterns, motions, depths, colors, etc. - just to mention a few. Due to this, visual data is

1. information-rich (i.e. it contains far more data than it can be analyzed by a practical vision system in real-time),

2. highly redundant,

3. and episodic.

Episodic means that visual events usually occur as bursts, i.e. they tend to be clumped in space (objects of interest) and time (events of interest). Therefore, computer vision algorithms require selective processing of regions of interest, since the camera's sensing range is usually lower than the area under observation.

To achieve selective sensing over space and time (e.g. to allow for wide-area target acquisition or multi-view computer vision algorithms such as stereo or 3-D reconstruction), cooperative gaze control of the participating cameras is necessary. Gaze control in its most general form is the alteration of imaging parameters to aid in the performance of visual tasks. These include the six degrees of freedom for camera position (x, y, z) and orientation (α, β, γ), see Section 2.3.1. The primary goal of gaze control is to actively manipulate the imaging system in order to acquire images which are well-suited to the tasks being performed. Due to this, the system is able to select a working domain in terms of image acquisition on its own, where it may achieve its highest performance for the underlying task, i.e. the image interpretation step. As depicted in Figure 2.1, we define a camera as an *Active Camera*, if it is able to manipulate the position control step and analyze the image data on-board. The process of image interpretation is implemented by using the concept of a Smart Camera [17], i.e. integrating processing capabilities into the sensor.

2.1. DEFINITION: ACTIVE CAMERA 25

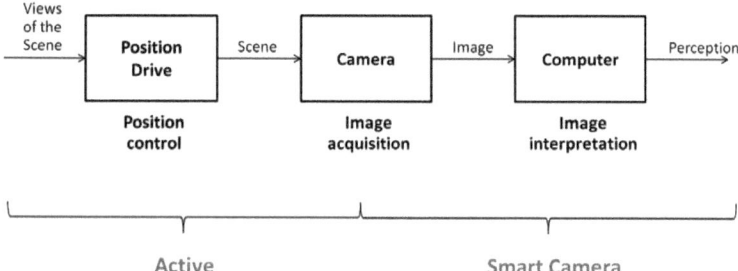

Figure 2.1: Active Camera concept

In this thesis, we will develop dynamic reconfiguration methods that are able to control the camera's location, i.e. based on the manipulation of position drives, in order to aid in the performance of visual tasks. The Active Camera's location is reconfigurable in terms of position (x, y, z) and orientation (α, β, γ) (see Section 2.3.1), whereas β and γ are assumed to be fixed in this thesis. By changing the position on demand, Active Cameras are able to meet the requirements for computer vision algorithms given through an imaging quality function, which depends on the underlying application. The mobile entity could be an unmanned vehicle, e.g. a mobile robot or an unmanned ground/air/underwater vehicle as described in Section 2.3.2. By using the concept of Smart Cameras for image interpretation, each Active Camera contains a computing unit in order to carry out image analysis and handle the organization of the mobile entity as well as cooperative tasks. An overview of existing Smart Camera prototypes is given in Section 2.4.1. Based on these components, reconfigurable Active Camera Networks can be built, which belong to this thesis.

The following list summarizes several of the advantages of Active Cameras [16]:

- **Overcome limited field of view / occlusions:** Any given camera system provides only a limited field of view of a scene. Active Camera Networks are able to overcome the restricted sensing range of one camera by repositioning it and can consequently capture new portions of a surveillance area. Additionally, repositioning of a camera often helps to overcome problems of occlusion.

- **Reduce computational complexity:** Using Active Camera Networks, it becomes possible to shift the system's attention to areas of activity. Thus, tracking times of objects can be increased. E.g. this may alleviate the process of handovers of objects between neighboring cameras by reducing the computational complexity of object association.

- **Multi-view algorithms:** Active Cameras can align themselves through movements in order to make way for multi-view computer vision algorithms, e.g. stereo or 3-D reconstruction. Additionally, controlling the aperture, zoom, and shutter speed (or lighting) can be used to manipulate the depth of field. E.g. large depth of field alleviates the use of stereo-based algorithms, whereas shallow depth of field increases the accuracy of object segmentation or detection.

In the following three sections, we explain from a technical perspective how targets can be detected in a surveillance area of a camera network and how the position control (including the image acquisition) and the image interpretation step can be realized.

2.2 Detecting Targets

In recent years, the integration of other low-cost sensor typologies is particularly interesting and promising for coping with issues and limitations emerging in camera networks. Thus, these low-cost sensors can be used for the fast and accurate detection of potential observation targets, which are fed into a high-cost camera system for visual observation. Thereby, the camera resources can be planned and used more efficiently.

Especially, for the detection of targets within a camera network's workspace, three-dimensional proximity measurement sensors can be used delivering the range, bearing, and elevation to a target from a known sensor pose, i.e. a

2.2. DETECTING TARGETS

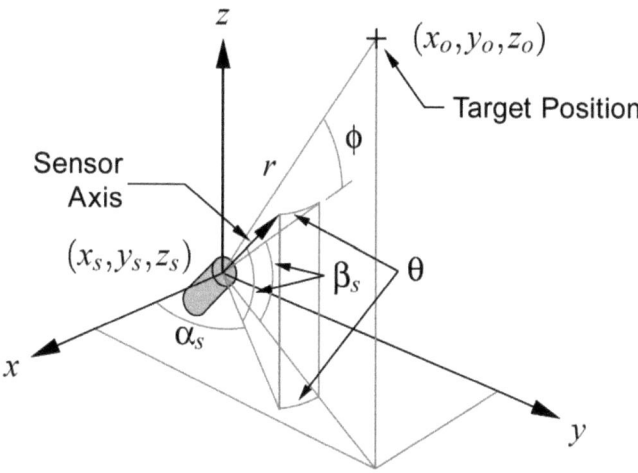

Figure 2.2: Measurement variables for a proximity sensor [1]

known sensor position and orientation. As depicted in Figure 2.2, the range r is the linear distance between the target and the sensor frame. The angular difference between the orientation of the sensor's axis in relation to the $x - z$ plane is α_s and θ in relation to the target's position, respectively. The elevation ϕ is the angular difference between the orientation of the sensor's axis with respect to the $x - y$ plane, β_s, and the target's position. Thus, the Cartesian position of a target relative to the world coordinate frame can be determined as follows:

$$x_0 = x_s + r\cos(\theta + \alpha_s)\cos(\phi + \beta_s) \qquad (2.1)$$

$$y_0 = y_s + r\sin(\theta + \alpha_s)\cos(\phi + \beta_s) \qquad (2.2)$$

$$z_0 = z_s + r\sin(\phi + \beta_s) \qquad (2.3)$$

Various non-contact sensors are available, which can be used to measure the proximity of a target. Generally, they are based on laser-triangulation, phase- or amplitude-modulation based electro-optical transducers or ultrasonic transducers [18]. A pair of calibrated CCD cameras can also be used in combination with stereo image-processing techniques to estimate the distance between a target and a center point between the cameras [19].

Utilizing data fusion, multi-modal sensor data stemming from multiple, imprecise sensors, e.g. motion detectors, can be combined or refined by map data (e.g. from the area of deployment) to create position estimations of targets at lower costs [20]. Data fusion methodologies, which are suitable for this kind of parameter estimation, contain the least squares estimator (LS) or similar approaches. Uncertainties of the measurement process can be considered through an ellipsoidal volume. The Kalman Filter (KF), and Extended Kalman Filter (EKF) for non-linear systems, are additional popular techniques for the fusion of multisensor data [21]. Kalman filtering is especially suited for real-time applications due to its iterative approach.

Based on the output of these sensors, cameras are able to plan their next position more efficiently. In the following section, we describe how the position control and image acquisition can be realized technically.

2.3 Position Control and Image Acquisition

Position control and image acquisition encompass mechanisms for controlling the camera's location in terms of its orientation and position. Therefore, hardware beyond the visual sensor itself is required to allow for this control. In this context, the term *control* refers to be the change of the viewing direction - called *sensor control* - as introduced in Section 2.3.1 or the Active Camera's repositioning as considered in Section 2.3.2. For both, mechanical hardware is needed, e.g. a motor controlled pan/tilt head to point the camera's head or a mobile entity to change the camera's position.

2.3. POSITION CONTROL AND IMAGE ACQUISITION

2.3.1 Sensor Control

Regarding the camera's orientation, the three orientational degrees of freedom (α, β, γ) (α = camera's viewing direction and angle of rotation around the vertical axis, β = angle of rotation around the horizontal axis, γ = angle of rotation around the optical axis) must be controlled. Nevertheless, this thesis considers only the camera's viewing direction α in terms of sensor control. These parameters are important to point a camera at features of interest. The basic mechanical properties of sensor control are listed below. An ideal sensor control platform could allow for actively adjusting each of these parameters. Nevertheless, the more parameters can be controlled, the more complexity is introduced into the system. Examples of these parameters are given as follows:

- For each camera:
    - pan (viewing direction α and angle of rotation around the vertical axis)
    - tilt (β, i.e. angle of rotation around the horizontal axis)
    - roll (γ, i.e. angle of rotation around the optical axis)
    - focus
    - aperture
    - zoom
    - optic axis calibration adjustments
- For multi-head platforms:
    - baseline

Certainly, the following characteristics can be considered as the most important aspects on the list: pan, tilt, aperture, and focus [16]. In addition to the mechanical features listed above, a sensor control platform should try to minimize its power consumption. The availability of commercial mechanical and optical hardware suitable for building Active Camera Networks has improved significantly over the last years[1]. In case of a network camera, the camera parameters can be computer controlled by establishing a network connection.

2.3.2 Examples for Mobile Entities

Concerning activity control through mobile entities, the Active Camera's positional degrees of freedom, (e.g. x, y, z), become controllable. In terms of inter-camera control, the baseline of neighboring cameras can be manipulated. This is important for multi-view computer vision algorithms such as stereo or 3-D reconstruction.

Unmanned Ground Vehicles Relating unmanned ground vehicles, we will present two autonomous, self-driving cars which participated in real traffic in 2010:

- Caroline from the University of Braunschweig
- Google car

Caroline is a standard 2006 *Volkswagen Passat* station wagon equipped with a variety of sensors, actuators, and computers to function as an autonomous mobile robot [2]. In its front, two multilevel laser scanners, one multibeam LIDAR sensor, and one radar sensor cover a field of view up to 200 meters for approaching traffic or stationary obstacles. In addition, four cameras detect and track lane markings in order to allow for precise lane keeping. The stereo vision system behind the windshield and another color camera combined with two laser scanners mounted on the roof were installed to provide information about the terrain in front of the vehicle. Very similar to the front of the vehicle, one multilevel laser scanner, one-medium-range radar, one LIDAR, and two radar-based blind spot detectors enable Caroline to detect obstacles at the rear. All these sensors are depicted in Figure 2.3. The sensor

---

[1] E.g. the Axis PTZ 214 network camera offers pan/tilt/zoom capabilities for less than EUR 1,000.

2.3. POSITION CONTROL AND IMAGE ACQUISITION

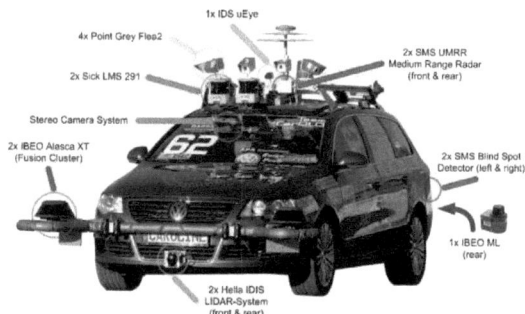

Figure 2.3: Caroline - Autonomous car from the University of Braunschweig [2]

data is processed locally and only based on this data the car decides how to drive.

In contrast to Caroline, the self-driving car from Google is based on a *Toyota Prius* and uses map data in addition to sensor data, which are provided by the Google data centers. Nevertheless, the Google cars drove 140,000 miles with occasional human interventions and 1,000 miles fully autonomously [22].

Both of the aforementioned cars could serve as mobile entities for Active Cameras.

Unmanned Air Vehicles With respect to activity control, technology from the field of unmanned air vehicles (UAVs) becomes more and more relevant. In the recent past, computer controlled quadro- and octocopters emerged, which can be navigated through mobile phones. A quadrocopter called *AR Drone* from the company *Parrot A.S.* can be controlled wirelessly and costs less than EUR 500 (depicted in Figure 2.4). Nevertheless, it is restricted in terms of its load-bearing capacity (maximum $250\,g$).

In addition to the aforementioned low-cost UAV, various companies (e.g. *Cassidian*) offer UAVs varying in terms of their size and weight ranging from kilograms to tons, which are less restricted in terms of their load-bearing ca-

Figure 2.4: AR drone from Parrot A.S. - controllable through Wi-Fi

pacity. In the research project *DEMUEBP*, *Cassidian Air Systems*[2] (formerly part of *EADS*) uses several UAVs in order to investigate their applicability for police missions.

UAVs could serve as a mobile entity for Active Cameras. In comparison to unmanned ground vehicles, aerial navigation is less complicated due to a lower frequency of obstacles in the actuation range. However, the cost of UAVs with acceptable load-bearing capacities is higher than for unmanned ground vehicles. Additionally, in order to operate outside "restricted" airspace, cumbersome procedures and regulations have to be passed, since it is critical that UAVs do not endanger other users of national airspace systems (e.g. commercial air traffic, cargo operations, and business jets) or compromise the safety of persons or property on the ground [23].

Unmanned Underwater Vehicles The first unmanned underwater vehicle (UUV) was developed at the Applied Physics Laboratory at the University of Washington as early as 1957 by Stan Murphy. Currently, UUVs as depicted in Figure 2.5, are used to operate autonomously in the ocean, e.g. for mine searching.

Hundreds of different UUVs have been designed over the past 50 years, but only a few companies sell a noticeable number of these vehicles. There are about 10 companies that sell UUVs on the international market, including

---

[2] http://www.cassidian.com

Figure 2.5: Bluefin-12 UUV with a Buried Object Scanning Sonar (BOSS) integrated in two wings

Kongsberg Maritime, *Hydroid* (now owned by *Kongsberg*), *Bluefin Robotics*, *International Submarine Engineering Ltd.* and *Hafmynd*.

Vehicles range in size from man portable lightweight UUVs to large diameter vehicles of over 10 meters length. Once popular amongst the military and commercial sectors, the smaller vehicles are now losing popularity. It has been widely accepted by commercial organizations that to achieve the ranges and endurances required to optimize the efficiencies of operating UUVs, a larger vehicle is required. However, smaller, lightweight and less expensive UUVs are still common as a budget option for universities. These smaller vehicles could be used as mobile entities for Active Cameras. Nevertheless, the underwater world is undoubtedly a difficult and challenging environment for computer vision algorithms due to the restricted sensing range including diffuse lighting.

2.4 Image Interpretation

Based on the quality of the image acquisition step, which can be improved by actively controlling the camera's position, image interpretation makes way for extracting information of the captured images. For image interpretation, the concept of Smart Cameras is chosen. Thereby, each Active Camera contains a computing unit in order to carry out image analysis and handle the organization of the mobile entity as well as cooperative tasks. An overview of existing Smart Camera prototypes is given in Section 2.4.1. Section 2.4.2 introduces single- and multi-camera computer vision algorithms for image interpretation.

2.4.1 Smart Camera Prototypes

Since the 1990s, Smart Cameras have attracted significant interest from research groups, universities, and many industry segments especially in video surveillance and manufacturing industries. The reason is that they offer distinct advantages over normal (or standard) cameras by performing not just image capturing but also image analysis and event/pattern recognition, all in one compact system. The growing popularity can be attributed to the progress made in semi-conductor process technology and embedded computer vision techniques, along with socio-economic factors such as the society's need for safety and security. Due to this reason, Smart Cameras are well-suited for image interpretation on Active Cameras. In addition to their powerful sensing, processing and communication units, they offer off-the-shelf interfaces like RS232 or USB to connect mechanical hardware like pan/tilt/zoom drives.

A generic architecture for a Smart Camera is depicted in Figure 2.6 consisting of a sensing, processing, and communication unit [3]. The concept of Smart Cameras, i.e. distributing the sensing, processing and communication resources throughout the camera network, enables the creation of scalable solutions in terms of image interpretation [24].

The image sensor may be integrated in either CMOS or high-resolution CCD technology and represents the data source of the processing pipeline. The raw data from the image sensor is read and pre-processed by the sensing unit. The pre-processing often comprises standard computer vision algorithms such as white balance or color transformations. The sensing unit also controls the setting of the image sensor with parameters such as the sampling rate. The main image processing tasks are performed at the processing unit. Here,

2.4. IMAGE INTERPRETATION

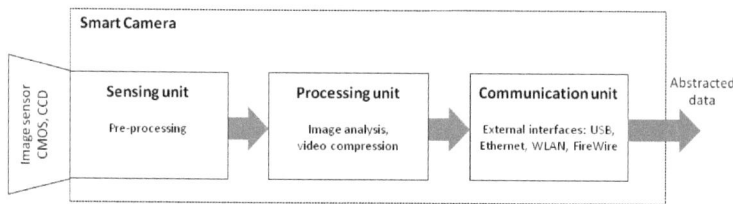

Figure 2.6: Generic architecture of a Smart Camera [3]

computer vision algorithms are fed with images delivered by the sensing unit. The output consists of abstracted data that is transfered to the communication unit, which provides various external interfaces like USB, Ethernet, WLAN, or FireWire. This abstracted data may be used for various purposes. In the most basic form, it is delivered to a human user and utilized for further evaluation, e.g. in case of a surveillance scenario. Nevertheless, the data may also be used for activity control without human intervention, e.g. triggering event-based reconfigurations. Thus, effective closed-loop systems can be created.

Various prototypes have been built for different fields of applications. A number of research projects focused on the problem of building high performance, stationary Smart Cameras for performing computing-intensive computer vision algorithms. These systems are usually based on popular computing platforms such as field programmable gate arrays (FPGAs), digital signal processors, and/or micro-processors. Due to their high demand for energy, they are stationary and connected directly to the mains. Recently, a new research direction came up focusing on building wireless Smart Cameras with very small size, at low cost and with low power consumption. The remainder of this section presents prototypes of wireless Smart Cameras, since they are best-suited for building Active Cameras.

Wireless Smart Cameras In order to increase the flexibility and mobility of camera networks, wireless communication has become an important design feature. Table 2.1 presents an overview of selected Smart Cameras using wireless communication and so being able to run in battery mode.

Table 2.1: Examples of wireless Smart Camera prototypes

System	Sensor	CPU	Communication	Power	Application
Meerkats [25]	Webcam 640x480 pixels	StrongARM at 400 MHz	802.11b	Battery	Local image analysis, Collaborative object tracking
Cyclops [26]	Color CMOS, 352x288 pixels	ATmega128 at 7.3 MHz	None on-board (802.15.4 via MicaZ mote)	Battery	Collaborative object tracking
MeshEye [27]	2 x low resolution sensor, 1 x VGA color CMOS sensor	ARM7 at 55 MHz	802.15.4	Battery	Unknown
WiCa [28]	2 x color CMOS sensor, 640x480 pixels	Xetal 3D (SIMD)	802.15.4	Battery	Local processing, Collaborative reasoning
CMUcam3 [29]	Color CMOS, 352x288 pixels	ARM at 60 MHz	None on-board (802.15.4 via FireFly mote)	Battery	Local image analysis, Inter-node collaboration
CITRIC [30]	OV9655 color CMOS sensor, 1280x1024 pixels	XScale PXA270	802.15.4	Battery (482-970 mW)	Compression, Tracking, Localization

2.4. IMAGE INTERPRETATION

The processing unit of *Meerkats* [25] is based on an Intel Stargate mote which is equipped with a 400 MHz StrongARM processor, 64 $MByte$ SDRAM, and 32 $MByte$ Flash. An embedded Linux system serves as operating system. Wireless communication is realized by an 802.11b standard PCMCIA card. The sensing unit consists of a consumer USB webcam with 640x480 pixels. Objective of the development of this prototype was to evaluate the power consumption of different tasks such as Flash memory access, image acquisition, wireless communication, and image data processing. In order to detect moving objects, the image data is processed locally on the camera itself. A master-slave mechanism is implemented to allow for collaboration between nodes, e.g. collaborative object tracking. Furthermore, the image data is compressed and transmitted to a central sink for potential post-processing.

In [26], Rahimi et al. present a Smart Camera called *Cyclops*. It is equipped with a low-performance ATmega128 8 Bit RISC microcontroller with 7.3 MHz and 4 $kByte$ of on-chip SRAM, and 60 $kByte$ of external RAM. The sensing unit delivers 24 Bit RGB images at CIF resolution (352x288 pixels). The communication unit is not on-board but can be attached by means of a MicaZ mote. A group of Cyclops cameras is used to implement an object tracking application.

At Stanford Wireless Sensor Networks Laboratory, Aghajan et al. [27] developed a Smart Camera prototype called *MeshEye* mote. The processing unit consists of an Atmel AT91SAM7S controller, which incorporates an ARM-7TDMI Thumb processor. The 32 Bit RISC processor can be clocked up to 55 MHz. The sensing unit consists of three image sensors. Two of them are low quality sensors acquiring images of 30x30 pixels with 6 Bit grayscale depth. The third one is a CMOS sensor with a VGA resolution, i.e. 640x480 pixels. There are eight additional slots for the integration of low quality sensors in order to observe a greater part of the surrounding area. Once an object has been detected, a second low-resolution sensor is activated and the location of the detected object is estimated using stereo vision. Then, the VGA sensor is activated in order to capture high quality images. The communication unit contains an IEEE 802.15 (ZigBee) device and is connected to the main processor's USB hub. The main advantage of this approach is that power consumption can be kept at a minimum as long as there are no objects in the camera's field of view.

Kleihorst et al. developed the *WiCa* wireless Smart Camera [28]. This is equipped with a SIMD processor IC3D operating at $80\,MHz$. This processor contains 320 RISC processing units, which are able to operate on the captured image data in parallel. The image data is stored in line memory. A 8051 microcontroller is available for general-purpose computations and communication tasks. The communication interface consists of an 802.15.4 device. The WiCa platform aims at low power consumption and is able to run in battery mode. Four WiCa Smart Cameras are used to build an application for distributed gesture recognition, i.e. collaborative reasoning.

In [29], a Smart Camera called *CMUcam3* was developed. It consists of a color CMOS sensor, which is capable of delivering 50 frames per second at a resolution of 352x288 pixels. The processing unit is an ARM7 microcontroller and operates at $60\,MHz$. The CMUcam3 is equipped with $64\,kByte$ of RAM and $128\,kByte$ of Flash memory. It contains a software layer with ready-to-use vision algorithms such as color tracking, frame differencing, and image compression. The communication unit is an 802.15.4 device, which can be attached to the serial communication channel. An application was set up for home activity monitoring.

The *CITRIC* Smart Camera [30] is a wireless camera hardware platform with an SXGA OmniVision CMOS sensor as sensing unit. The processing unit is an XScale processor equipped with $64\,MByte$ of RAM and $16\,MByte$ of Flash memory. An 802.15.4 device is connected to the CITRIC board. The CITRIC prototype has been demonstrated in image compression and applications such as single-target tracking via background subtraction and camera localization using multi-target tracking.

As can be seen from the aforementioned systems, several prototypes that operate as nodes in wireless Smart Camera networks have been proposed and tested in real-world scenarios. Wireless Smart Cameras deliver the flexibility and mobility needed to create Active Cameras as considered in this thesis. Especially, the *Meerkats* and the *CITRIC* Smart Camera are suitable to build Active Smart Camera networks, since they possess enough computational power

2.4. IMAGE INTERPRETATION

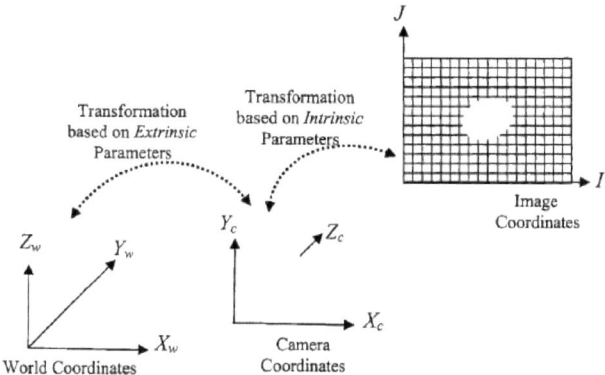

Figure 2.7: Image, camera, and world coordinate frames [4]

to handle sophisticated computer vision algorithms and capture images with a resolution of up to 1,280x1,024 pixels. Furthermore, standard image processing libraries such as Intel's *OpenCV* (Open Source Computer Vision) [31] can easily be ported to these cameras. Additionally, they are equipped with good communication abilities, e.g. the *Meerkats* Smart Camera uses IEEE 802.11 WLAN. Thus, enough bandwidth is available for fast and high-load message exchange between cameras collaborating on a surveillance task. The setting of both cameras is very similar to the technical setting of the quadrocopter depicted in Figure 2.4. Thus, code developed for both Smart Cameras could be ported and run on-board on the quadrocopter.

2.4.2 Computer Vision

Since image interpretation based on computer vision is an integral part of Active Camera Networks, we present a short overview about single- and multi-camera algorithms in this section. This is to provide a deeper understanding of the calculus caused by computer vision algorithms.

Single-Camera Algorithms Single- and multi-camera computer vision algorithms are usually based on the assumption that cameras are calibrated. The basic idea behind *camera calibration* is to calculate real world distances from an image that has been captured by an image sensor, see Figure 2.7. The calibration process includes two steps. First, the intrinsic parameters such as focal length are estimated to allow for transforming the image coordinate system into the camera coordinate system. Secondly, the estimation of the extrinsic parameters such as the camera's position, translation and rotation allow for the reconstruction of the world coordinates from the camera coordinates. Various methods and algorithms have been proposed for camera calibration. The method presented by Tsai [32] is useful for estimating the camera's external position and orientation relative to the object's reference coordinate system as well as the focal length, radial lens distortion, and image sensing parameters. Cheng et al. [33] investigated a method for obtaining vision graphs for distributed Smart Camera networks to make way for network calibration. The vision graph allows to determine, which cameras share overlapping fields of view. This is especially important in case of system-wide object tracking.

In order to detect objects and their movements, methods of *segmentation and motion detection* are used. A common approach is background subtraction. Toyama, Krumm, Brumitt, and Meyers give a good overview and comparison of many techniques [34]. In order to perform background subtraction, a model of the background must be "learned". Once learned, objects can be separated from the background by computing the difference between the current frame and the background of the scene. The objects left after subtraction are presumably new foreground objects. Although background modeling methods work fairly well for simple scenes, they suffer from an assumption that is often violated, i.e. that all the pixels are independent. This assumption does not hold due to the fact that the brightness of pixels depends on their neighbors. Therefore, sophisticated models exist to take this into account. However, since these models come at extra cost in terms of memory consumption and computational effort, they are usually avoided and the simple background subtraction model is used in practice. In the case of moving

2.4. IMAGE INTERPRETATION

cameras, optical-flow-based algorithms [35] are used for dynamic background subtraction, e.g. for obstacle avoidance in cluttered environments [36]. Optical flow is the pattern of apparent motion of objects, surfaces, and edges in a visual scene caused by the relative motion between an observer (an eye or a camera) and the scene. The concept of optical flow was first studied in the 1940s and ultimately published by American psychologist James J. Gibson as part of his theory of affordance. Optical flow techniques such as motion detection, object segmentation, time-to-collision and focus of expansion calculations, motion compensated encoding, and stereo disparity measurement utilize this motion of the objects, surfaces, and edges. In Section 5.3.4 of this thesis, the concept of optical flow is used to detect the offset of feature points over time to compute so-called *saliency curves* for collaborative time synchronization between neighboring cameras.

Object detection is an important capability of Smart Cameras. Detecting objects means to identify their positions in an image and to identify the object itself. Object detection has matured over the last years and is now commercially available. Even standard digital cameras for the consumer market offer sophisticated object detection algorithms such as face detection[3]. A very promising approach for object detection was developed by Viola and Jones [37] and is based on so-called *Haar-like* features. These features are sums of pixels in rectangular areas and can be used to construct decision trees (called classifier cascades) encoding sophisticated object shapes.

Based on the output of single-camera algorithms, multi-camera algorithms can be set up for network-wide object association and tracking allowing for cooperative image interpretation.

Multi-Camera Algorithms *Object association and tracking* across multiple cameras is important due to the limited fields of view of single cameras and object occlusions in real scenes. Therefore, single cameras are not able to completely observe an area of interest on their own. Instead, multiple cameras are used to widen the active area of observation. To enable cooperative sensing among multiple cameras, objects have to be recognized among them.

---

[3]E.g. Casio EXILIM EX-Z1

This is done by the process of object association. Hereby, a more coherent understanding about what is occurring in the observed scene can be constructed. Several approaches with varying constraints have been proposed. For instance, the problem of associating objects across multiple stationary Smart Cameras with overlapping fields of view has been addressed in a number of tracking applications, e.g. [38, 39]. If the problem of object association is extended to cameras with non-overlapping fields of view, geometric and appearance-based approaches are used [40, 41]. Usually, this is based on trying to associate data of the object, e.g. its color histogram, feature points etc., among the participating cameras. Camera motion has also been studied where correspondence of pixels is estimated across pan/tilt/zoom cameras [42]. Modern approaches try to use machine learning techniques, such as neural networks, in order to improve the assignment of actual objects to tracked objects in terms of object association, e.g. [43].

The problem of shape reconstruction from pairs of images (e.g. of neighboring cameras) is known as *stereo vision*, which is one of the oldest in computer vision [44]. In order to achieve this, the correspondence problem has to be solved, i.e. finding areas of each image corresponding to the same point in the scene. If such a correspondence has been found, the point can be triangulated to determine its coordinates in three dimensions. The corresponding output is a depth map computed from all the correspondences found between the images. For static scenes, approaches known as dense stereo and feature matching are used. Since Active Cameras are able to move, dynamic stereo can be used. Dynamic stereo uses motion cues in the image to aid in the depth map construction, either from a dynamic scene or from a moving camera [45].

2.5 Summary

This chapter contained a description of the basic components of Active Camera Networks.

First, we explained how the concept of Active Cameras can help to alleviate computer vision in terms of image acquisition and interpretation. Afterwards,

2.5. SUMMARY

hardware platforms for Active Cameras have been introduced to give an insight in how far reconfiguration can be used and which prerequisites need to be met. In terms of image interpretation, the concept of Smart Cameras was introduced to extract image information near the image sensor. Extracting these information can be achieved by a number of computer vision algorithms, which have been presented at the end of the chapter.

We concluded that there are sophisticated Smart Camera prototypes and mobile entities, such as unmanned ground/air/underwater vehicles, which can be used to build Active Camera Networks. Since they are based on off-the-shelf hardware, e.g. ARM-based chipsets, standard image processing libraries such as Intel's *OpenCV* (Open Source Computer Vision) can easily be ported to these platforms. This alleviates the process of developing image processing algorithms for these devices. In addition, mobile entities - such as the *AR Parrot's* quadrocopter - are equipped with powerful on-board communication and computing resources. Thus, the Smart Camera concept can be integrated into these entities. Furthermore, the available communication abilities make way for enough bandwidth within the network for fast and high-load message exchange between cameras collaborating on surveillance tasks.

Based on this outcome, the following chapter introduces the system model and explains which assumptions are made to model Active Camera Networks as they have been presented in this chapter.

Chapter 3

System Model

Visual surveillance of targets deals with image acquisition and interpretation of captured data of moving targets. In this thesis, a set of Active Cameras cooperates with a set of so-called *perceiver nodes* generating *target requests*. Perceiver nodes are low-cost sensors acquiring proximity measurements of targets by appropriate sensor technology at specific demand instants. Since they are low-cost, they can be deployed on the surveillance area in high numbers. Active Cameras react on requests to the perceiver nodes in their actuation range to acquire high-quality target imageries. In order to achieve this, they exchange information about their current state (alignment, position of targets, etc.). In the remainder of this chapter, the components of such a system are described in detail.

3.1 Active Cameras

Each Active Camera is an autonomous node containing a mobile entity (e.g. a quadrocopter as depicted in Figure 2.4) and a camera with integrated processing capabilities (CPU, memory etc.) and a communication interface, i.e. a Smart Camera. They are initially deployed on a spatial area, the Active Camera Network's surveillance area. The use of broadcast communication allows for an efficient usage of the wireless communication channel and enables the Active Cameras to establish local neighborhoods. Broadcast communication is realized on layer two of the protocol stack, where packets are not repeated upon communication failures, sent within a bounded randomization interval, and are queued in a finite interface queue. Local neighborhoods carry out tasks cooperatively. The network connecting Active Cameras is a wireless mobile ad-hoc network (MANet), since this allows for fast and dynamic reconfiguration. In such a MANet, we assume the transmission range of an Active Camera to be at least two times its actuation range to avoid network partitioning.

The region of the surveillance area that an Active Camera can travel to is referred herein as *actuation range*, see Figure 3.1. The Active Camera's actuation range is defined by its actuation radius and center of movement, whereas the center of movement is the camera's position of deployment. The actuation range defines the camera's reachable region by repositioning, see Section 3.1.2. Local neighborhoods resulting from spontaneously connected cameras in sending/receiving range do not posses any knowledge beyond their own communication range. Active Cameras communicate with neighboring nodes via a wireless communication channel. For the used communication technology, it is assumed that the delay jitter δ_{lat} is bounded. This can, for example, be achieved by using broadcast communication on layer two of the protocol stack as explained before. Thus, packets are not repeated upon communication failures, sent within a bounded randomization interval, and are queued in a finite interface queue. The assumption of a bounded delay jitter will be used to prove the correctness, see Section 5.2.5, of our reconfiguration algorithm introduced in Section 5.2.

Each Active Camera has information about its current position. This may be obtained by appropriate technologies such as GPS (or Galileo that should be operational by 2013) in outdoor scenarios or by IEEE 802.11 LAN positioning in indoor scenarios [46]. Further, it is assumed that the coordinate system is a metric space that allows to calculate the Euclidean distance between any given pair of cameras. Additionally, we assume all cameras are calibrated, see Section 2.4.2. Thereby the position and size of targets can be derived from

3.1. ACTIVE CAMERAS

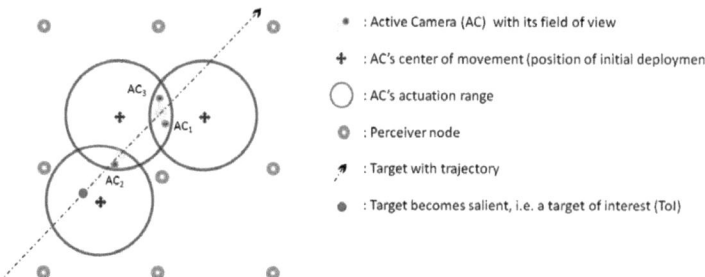

Figure 3.1: System model: Active Cameras, perceiver nodes, and dynamic targets

images captured by the camera. Cameras are able to focus on a target of a certain size at a particular world location. Typically, the 3-D location of a target is its center point. E.g. for biometrics, this would be the center point of the face.

An Active Camera prototype can be based on a quadrocopter as depicted in Figure 2.4. It contains an ARM9 processor with 468 MHz, 128 $MByte$ of DDR RAM and Wi-Fi b/g. The frontal camera is able to capture images of 640x480 pixels and the device is running Linux. As image processing library Intel's *OpenCV* (Open Source Computer Vision) can be chosen, see [31], since this is a widely-used open source project with a high reputation in the computer vision community and can be easily cross-compiled for the ARM chipset.

3.1.1 Field of View

As depicted in Figure 3.2, we assume the camera's field of view and its geometry can be simplified according to De Floriani's definition [5]. The algorithms proposed in this thesis rely on 2-dimensional geometries, i.e. $z = 0$. It can be extended to a 3 dimensional model, which would make way for a more sophisticated model for dynamic reconfiguration. Nevertheless, the computation of volumetric fields of view does in return require a much higher computational

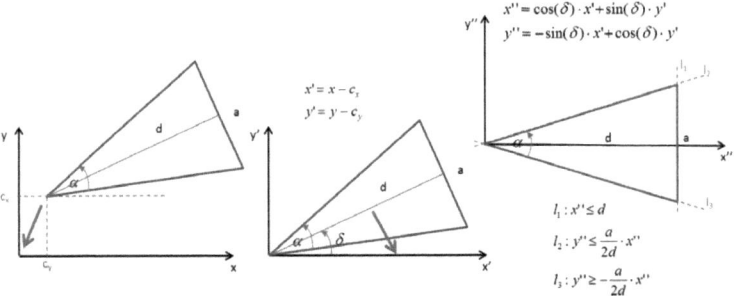

Figure 3.2: Geometry of an Active Camera's field of view [5]

effort [47]. Due to this reason, this has not been investigated further in context of this thesis.

In our case, the node's ground plane view is approximated by a triangular shape. Although real world experiments show that it is rather similar to a trapezoid, this simplification has only minor impact on the accuracy for alignment and positioning. In [48, 33], for example, the camera's field of view is modeled as a triangular shape serving as a basis for the implementation of a positioning and calibration algorithm. Thus, an Active Camera's field of view is characterized by its pose, i.e. location (x, y), orientation δ, span angle α, and the distance d (determined by the camera's focal length) as described in Figure 3.2. The following three linear constraints define the area covered by the Active Camera's field of view and can be used to check whether a dynamic target with position (x, y) can be seen or not:

$$\cos(\delta)(x - c_x) + \sin(\delta)(y - c_y) \leq d \tag{3.1}$$

$$-\sin(\delta)(x - c_x) + \cos(\delta)(y - c_y) \leq \frac{a}{2d}[\cos(\delta)(x - c_x) + \sin(\delta)(y - c_y)] \tag{3.2}$$

3.1. ACTIVE CAMERAS

$$-\sin(\delta)(x-c_x)+\cos(\delta)(y-c_y) \geq -\frac{a}{2d}\left[\cos(\delta)(x-c_x)+\sin(\delta)(y-c_y)\right] \quad (3.3)$$

3.1.2 Camera's State

We assume that Active Cameras are installed in a way that collisions are avoided during reconfiguration in case of overlapping actuation ranges. Furthermore, they know about obstacles initially (preconfigured by maps) or from the images they collect and analyze. Due to the distributed architecture we use, each camera only needs to know about obstacles in its own sensing range and no global knowledge of all obstacles in the system is required. For the detection of obstacles, advanced techniques need to be used.

The Active Camera's state, $AC.state$, is represented by the following vector:

$$AC.state = (d,\ A_r,\ p) \quad (3.4)$$

Center of movement d is the AC's position of deployment.

Actuation range A_r is the AC's actuation range defining its reachable region by repositioning. The actuation range is influenced by the AC's actuation radius and its center of movement (circular area around d).

Current position $p = (x,\ y,\ z,\ \alpha)$ is the AC's current position $(x,\ y,\ z)$ in its actuation range including the orientation α.

3.1.3 Clock Synchronization

The majority of today's computer systems (including Active Cameras) are based on clocked circuits and hence contain hardware clocks. The cameras' local clocks are a valuable tool for visual surveillance, since they offer time stamps for visual events. Based on these time stamps, visual events can be ordered locally or fused across multiple cameras. A typical hardware clock consists of a crystal-stabilized oscillator and a counter that is incremented by one every oscillation period (e.g. upon detection of a falling or rising edge). Based on the periodic time T of the oscillator, the counter h can be used to obtain approximate measurements of real-time intervals in multiples of T. Thus, the clock counter has the value $h(t)$ at real time t and is incremented by one with a frequency of f. The rate of the counter is defined as $f(t) = dh(t)/dt$. An ideal clock would have rate of one at all times, but the rate of a real clock fluctuates over time due to changes in supply voltage or temperature - just to mention a few. Usually, the deviation of the rate from the standard rate 1 is assumed to be bounded. This deviation is called the clock's drift $\rho(t) = f(t) - 1 = dh(t)/dt - 1$, and denotes the corresponding bound with ρ_{max} [49].

$$-\rho_{max} \leq \rho(t) \leq \rho_{max} \quad \forall t \quad (3.5)$$

A reasonable additional assumption is $\rho(t) > -1$ for all times t. This means that a clock can never stop ($\rho(t) = -1$) or run backward ($\rho(t) < -1$). The oscillator's rate is given by the hardware manufacturer. Typically, Active Cameras contain off-the-shelf oscillators, and thus we have $\rho_{max} \in [1\text{ppm}, 10\text{ppm}]$[1] [49]. The frequency f_{sync} for re-synchronization in order to guarantee a clock deviation of less than δ seconds can be computed as follows: $f_{sync} = \rho_{max}/\delta$. In this thesis, we assume a bounded clock drift ρ_{max} of 10ppm and a maximum acceptable clock deviation of $\delta = 40\,ms$, i.e. one frame in case of frame rate of 25 fps. This means that its maximum drift is $10\mu s$ per second, which corresponds to 40ms in 4,000 seconds. Since we assume that the Active Cameras sample their environment with a frame rate of 25 fps, the Active Cameras should be synchronized with a frequency of $f_{sync} = 0.25\,mHz$. Therefore, this synchronization procedure has to be scheduled by the camera hourly as recovery task.

---
[1] Parts per million, that is 10^{-6}. A clock with a drift of 10 ppm drifts 10 seconds in a million seconds, or $10\mu s$ in one second.

3.2 Perceiver Nodes

The Active Cameras' task is to cooperatively observe targets of interest (ToIs), i.e. targets which have become salient on their way through the surveillance area, see Figure 3.1. Whether a target is salient or not depends on the surveillance scenario, e.g. on public places salient targets could be people running. To detect these ToIs, Active Cameras interact with *perceiver nodes*. Perceiver nodes could be security personnel, smart sensors to identify and quantify perceivable events in their vicinity, stationary camera-based systems used for object tracking or three-dimensional proximity measurement sensors as presented in Section 2.2. We assume that a perceiver node is deployed at each AC's center of movement with a sensing range defined as twice the camera's actuation range. Thus, they are distributed throughout the workspace to ensure sampling the entire target trajectory of moving targets. Moreover, they are able to identify and quantify occurring perceivable events in their vicinity.

We assume that perceiver nodes are three-dimensional proximity measurement sensors delivering the range, bearing, and elevation to a target from a known sensor pose, i.e. a known sensor position and orientation. In addition to the position, the time of the position estimation has to be added by the perceiver nodes. Therefore, we assume that the perceiver nodes are synchronized by means of traditional network synchronization like NTP [50]. Position estimations of targets within the perceiver nodes' ranges are delivered at a specific frequency, since target requests are sent with a frequency of $f_p Hz$. This is necessary to allow for data fusion within the network. Perceiver nodes deliver the position of targets containing a time stamp.

The next section explains how this data is used to generate target requests for the Active Camera Network.

3.2.1 Target Requests

If a target is characterized by a perceiver node as a target of interest (TOI) due to a salient behavior or by exceeding a pre-defined threshold of the sensor, a target request is generated. A unique identifier is generated for each target as soon as it enters the range of a perceiver node in the first place. It is assumed that perceiver nodes are able to maintain the target's identifier across demand instants locally and network-wide by estimating future positions, e.g. based on Kalman filtering. Thus, a target request TR consists of a target identifier ID, the location p (a three-dimensional vector including x, y, and z) and a time stamp t_{occ}:

$$TR = (ID,\ p,\ t_{\text{occ}}) \tag{3.6}$$

In this thesis, targets are assumed to be created dynamically during runtime rather than being given initially. This makes the system as independent as possible from *a priori* knowledge. The performance of the Active Camera Network depends on fulfilling its observation objectives (all target requests that are valid for a demand instant) in order to be able to capture the necessary imagery for the surveillance task. Target requests are generated by perceiver nodes and sent to the Active Camera Network through a dedicated communication channel.

3.2.2 Modeling Perceiver-Observation Uncertainty

Position estimations from perceiver nodes can contain errors. This may be due to unexpected changes or unmeasured variances in the perceiver itself. Since we assume that the perceiver nodes are calibrated, systematic errors stemming from calibration errors are neglected. Nevertheless, the sensor's observation can also be influenced by random errors. Random errors are the property of the perceiver and, thus, cannot be altered. Therefore, the perceiver's parameter estimation is modeled for a target's location by p_e, which is the correct value p corrupted by a random localization error modeled by a circle of radius h around p (see Section 6.2.1 for results).

3.3 Summary

In this chapter, our system model has been introduced, which will be used as a basis for the system architecture and dynamic reconfiguration methods in the following chapters.

Our system model consists of Active Cameras, which interact with low-cost perceiver nodes. Perceiver nodes are three-dimensional proximity measurement sensors delivering a location and time stamp of targets within the camera network's surveillance area. A perceiver node is deployed at each camera's center of movement with a sensing range defined as twice the camera's actuation range. In case of the detection of a target, a target request is generated by a perceiver node and sent to the camera network. Based on the concept of perceiver nodes and target requests, highly dynamic surveillance areas can be modeled.

Each Active Camera is an autonomous node containing a mobile entity (e.g. a quadrocopter) and a camera with integrated processing capabilities (CPU, memory, etc.) and a communication interface, i.e. a Smart Camera. Active Cameras are able to exchange messages with neighboring nodes. Thereby, local neighborhoods can be established. Local neighborhoods carry out tasks cooperatively. The network connecting Active Cameras is a wireless mobile ad-hoc network (MANet). In such a MANet, we assume the transmission range of an Active Camera to be at least two times its actuation range to avoid network partitioning.

Based on this system model, we present our system architecture in the following chapter.

Chapter 4

System Architecture for Active Cameras

The most important design feature of an Active Camera Network as introduced here is to support the reconfiguration of participating nodes in terms of their pose, i.e. orientation and position. The system architecture presented in the following allows for this reconfiguration and relies on a network of Active Cameras and perceiver nodes. Active Cameras need to reconfigure their pose in order to observe dynamic targets cooperatively and consider target requests sent by perceiver nodes, such as proximity sensors. This chapter presents basic functionalities concerning positioning and coordination of Active Cameras. This serves as a basis for the dynamic reconfiguration methods as introduced in Chapter 5. In Section 4.1, we discuss requirements for Active Camera Networks. Section 4.2 introduces our system architecture, which has been implemented as middleware. Finally, Section 4.3 concludes the chapter with a summary.

4.1 Requirements of Active Camera Networks

In comparison to traditional middleware implementations, a middleware for Active Camera Networks has to fulfill additional requirements. This is not merely due to different resource constraints but is also a consequence of the application domain [3]. Active Camera Networks are intended for processing captured images close to the sensor. This requires the support of sophisticated image processing algorithms through the middleware. In addition, wireless communication is relatively expensive compared to processing and is thus used sparingly (e.g. in case of certain events or to send aggregated sensor data to a base station). Collaboration of individual nodes is typically inherent to the application. Active Camera Networks demand higher communication bandwidth in order to exchange features extracted from images, or even streaming of the video data.

Typical Active Camera Network applications consist of tens to hundreds of Active Cameras (e.g. in wide-area scenarios such as airports or train stations), and various different tasks have to be executed. Assigning tasks manually to the Active Cameras is almost impossible. Due to this reason, a user should simply define the system goal through a set of tasks that have to be carried out (e.g. person tracking) combined with some restrictions and action rules to define the behavior in case of an event. The Active Camera Network itself then has to allocate the tasks to the Active Cameras. If a task cannot be performed by a single camera (e.g. creating a multi-view perspective of a scene), Active Cameras have to organize themselves and collaborate to fulfill the demands. This makes way for a self-organizing system, which should be based on a distributed and decentralized control in order to increase the robustness of the system.

The following list summarizes the requirements of Active Camera Networks:

- Support of sophisticated image processing algorithms

- Process and aggregate data before communication

- Simple definition of the system goal by the user

- Usage of distributed and decentralized control algorithms for scalability and robustness of the system

Our system architecture has been designed as middleware for Active Camera Networks. This is situated between the application and the underlying operating system, network protocol stack, and hardware. It aims at bridging the gap between application programs and the lower-level hardware and software infrastructure in order to ease the development of distributed systems [51]. The following section introduces our system architecture fulfilling the aforementioned requirements.

4.2 Adaptive Location Management Architecture

Given the previously introduced requirements for Active Camera Networks, it is obvious that a different kind of middleware is necessary. Middlewares for wireless sensor networks (cf. Section 7.2.2-7.2.3) are not intended to handle advanced image-processing tasks and sending large amounts of data. On the other hand, adapting a general-purpose middleware (cf. Section 7.2.1) is feasible and could fulfill the aforementioned requirements, although the introduced overhead does not yield efficient resource utilization. Therefore, we introduce a new software architecture for Active Camera Networks in the remainder of this section, which has been implemented as middleware for the ARM chipset, paving the way for scalability and adaptivity in these networks.

4.2.1 Architecture Overview

The reconfiguration methods presented in this thesis have been integrated into our middleware [52] that is presented in Figure 4.1. It shows a block diagram of the software components forming a single Active Camera. Its distributed

58 CHAPTER 4. SYSTEM ARCHITECTURE FOR ACTIVE CAMERAS

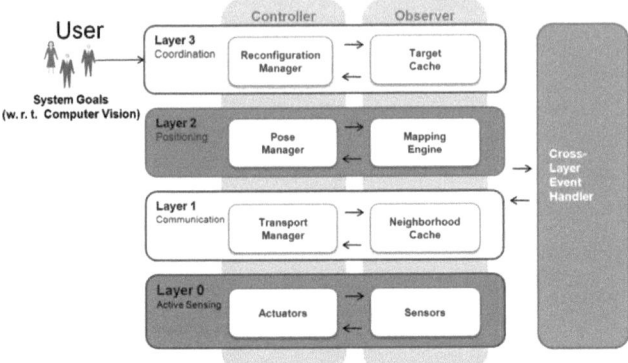

Figure 4.1: Single Active Camera: Adaptive Location Management Architecture

Observer/Controller design makes way for scalability and adaptivity in these networks. Our design consists of the following layers:

- Layer 0: This layer contains the basic functionalities for active sensing and processing such as the capturing and low-level analysis of image data and the physical reconfiguration of the camera's pose, i.e. the position and orientation. The low-level analysis of image data can be pre-processing of imagery such as white balance or color transformations, see Section 2.4.1.

- Layer 1: The layer is responsible for the communication and contains the *Transport Manager* and the *Neighborhood Cache*. The main purpose of the layer is to keep information about neighboring nodes, since their mobility may lead to frequent state changes.

- Layer 2: In this layer, the *Mapping Engine* and *Pose Manager* are included, which aim at determining the optimal pose of an Active Camera for visual surveillance of a corresponding target within the workspace. For this purpose, the camera's current location, motion capabilities, point in time of requested observation, and the model of the environ-

4.2. ADAPTIVE LOCATION MANAGEMENT ARCHITECTURE 59

ment are considered. In order to map the camera's position to real-world coordinates, a map of the camera's actuation range must be available. This could be given by the user initially or generated at runtime by sophisticated reconstruction mechanisms as presented by Sester et al. [53]. In case of a cluttered environment, the *Mapping Engine* would also be responsible to detect obstacles within the camera's actuation range. In order to learn the locations of obstacles, range sensors and similar techniques as presented in [54] could be used. Such a laser ranging device could be mounted on the Active Camera.

- Layer 3: The objective of this layer is to handle incoming requests and organize the execution of available reconfiguration methods in order to maximize the system's overall performance. In this layer, the dynamic reconfiguration methods of Chapter 5, i.e. *DRofACN* and *ACFSync*, are situated.

  Nevertheless, other methods could be added to this layer to be scheduled by the *Reconfiguration Manager*. In many surveillance applications, for example, it is of high interest to analyze spatio-temporal datasets for recurring patterns, e.g. to automatically extract pedestrian trajectories. In [55], Sester et al. present a centralized approach, which is able to examine spatio-temporal datasets for movement patterns. The key feature of the approach is that it is highly adaptive, i.e. patterns are not specified in advance (as meaningful semantic patterns) but established from the data set. Adding an algorithm to the *Reconfiguration Manager* gathering all relevant spatio-temporal data from neighboring nodes, would pave the way for utilizing such methods for distributed pattern recognition across the network. Nevertheless, this has not been investigated in this thesis.

- Cross-Layer Event Handler: Context data acquired by Active Cameras or other integrated sensors can be used through this component in every layer. Thereby, applications beyond the scope of classical wireless sensor networks become feasible. For instance, an optical compass (cf. [56]) or other activity recognition based algorithms [57], which would be situated in *Layer 2* or *Layer 3* but using visual data from *Layer 0*, could be realized.

The core layers making way for reconfiguration within Active Camera Networks are *Layer 2* and *Layer 3*. The coordination layer is responsible for determining how many and - even more important - which cameras are to be used for target observation in order to optimize the performance of the surveillance system for current and upcoming target requests. The positioning layer helps to determine the optimal pose of each node for active target requests.

The following sections give a detailed description of each layer.

4.2.2 Layer 0: Active Sensing

The *Actuators* component is implemented as standalone process able to control IP-based position drives. As one example, an interface for controlling the position drive of the Axis PTZ 214 camera [1] has been implemented. Currently, the pan and tilt angle, and the zoom setting can be controlled. Apart from the camera's PTZ abilities, control for other settings like built-in autofocus and white balance have been implemented. Future versions could contain software for controlling more sophisticated mobile entities, such as the *Parrot's AR Drone*[2], see Section 2.3.2. A socket-based communication scheme is used as unified interface.

For the *Sensors* component, an extensive software library for computer vision (*OpenCV* [31]) is used for image acquisition and interpretation. *OpenCV* contains various computer vision algorithms which incorporate many functions for image acquisition and interpretation as introduced in Section 2.4. Additionally, *OpenCV* is able to save still images, record video streams, and encode them in different ways. This data can be accessed by the *Cross-Layer Event Handler*, which is presented in Section 4.2.6, in order to forward them to subscriber components of other layers. For invocation and data transfer purposes, a socket based communication scheme is used. The exchange of image data

---
[1] http://www.axis.com/products/cam_214/
[2] http://ardrone.parrot.com

4.2. ADAPTIVE LOCATION MANAGEMENT ARCHITECTURE

between the sensor elements and other elements is done in shared memory. The shared memory block can be accessed from all components that need to work on image data.

4.2.3 Layer 1: Communication

As stated in Chapter 3, Active Cameras communicate via wireless ad-hoc networks. Network functionalities are of major importance for a self-organizing system.

Message exchange is coordinated by the *Transport Manager* which is a thread running on each Active Camera managing a message bag for outgoing messages. As soon as the thread is started, a timer is initialized. In case the timer expires, all messages that have been collected in the message bag are sent. Messages are sent using broadcast communication, since this allows for an efficient usage of the wireless communication channel and enables the Active Cameras to establish local neighborhoods. Broadcast communication is realized on layer two of the protocol stack, where packets are not repeated upon communication failures, sent within a bounded randomization interval, and are queued in a finite interface queue, see Section 3.1. In case the message bag was empty, the following heartbeat message is sent in order to inform neighboring nodes that the camera is still alive:

- Sender ID
- Message ID
- Time stamp
- Position (x,y,z)
- Orientation δ
- Span angle α
- Actuation radius
- Center of movement

The total length of a heartbeat message is 40 Byte, since all parameters contain integer values. Data from received heartbeat messages are stored in the *Neighborhood Cache* of each neighboring camera.

Algorithm 1 Heartbeat Algorithm

1: **init:**
2:   *set timer //timer for broadcasting heartbeat message, e.g. 2.5s*
3:   *set TTL //time to live, e.g. 10s*
4:   *set NC ← empty //start with empty neighborhood cache*
5:
6: **on** *incoming heartbeatMsg* :
7:   **if** NC contains information about neighbor already
8:     update information about neighbor in NC
9:   **else**
10:     add neighbor to NC
11:   **end if**
12:   set time-to-live of neighbor to TTL
13: **end on**
14:
15: **on** *timerexpire* :
16:   broadcast a heartbeat message
17:   delete neighbors with expired TTL
18: **end on**

The *Neighborhood Cache* component is a data structure. It collects various information relevant to the reconfiguration methods, such as the position of the neighboring nodes. The *Neighborhood Cache* can be extended to perform specific query and update tasks of its data for other elements. Algorithm 1 gives a short overview about the algorithm in pseudo code. Further background information about the implementation of the *Neighborhood Cache* can be found in [58].

4.2.4 Layer 2: Positioning

The objective of the *Positioning Layer* is to determine the optimal pose of a node for the surveillance of a particular target, i.e. to fulfill the sensing objective, which was given by the *User*, see Figure 4.1. For this purpose, the

4.2. ADAPTIVE LOCATION MANAGEMENT ARCHITECTURE 63

Active Camera's current location, its motion capabilities (e.g. whether it is a pan/tilt/zoom camera or a mobile robot), the time until the specific target request is obsolete, and the model of the environment have to be considered.

In order to fulfill the sensing objective, the various requirements of the computer vision algorithms which are responsible for the image interpretation step, see Section 2.4, have to be translated into concrete requirements for the positioning of the camera. An optimal position is achieved by maximizing the probability of successfully completing the addressed sensing objective, which is encapsulated into expected capture conditions, including distance at which a subject is imaged, angle of capture, and several others.

For a tracking application, for example, the sensing objective could be translated into requirements for the positioning as follows: People have to be captured from a side view in order to alleviate the process of background subtraction, since the object's motion is pronounced stronger from such a perspective.

In this thesis, we assume the sensing objective for visual surveillance to be frontal face detection. For frontal face detection, we assume the requirements of the algorithm presented in [6]. This algorithm requires that a face in the Active Camera Network's workspace has to be mapped to at least 20x20 pixels in the image, see Figure 4.2, and a view angle of less than 15 degrees. We assume that a human face has a size of 20x20 centimeters. That means that $1\,cm$ of the face has to be mapped to $1\,px$ in the image, i.e. $l_{res} = 0.01\,m$. Given the camera's angular field of view in horizontal direction α (e.g. 48° in case of an Axis PTZ 214 with a focal length of $4.1\,mm$), and the image size in pixels (e.g. 4CIF: 704x576 pixels, i.e. $w_{px} = 704$), we can compute the maximum distance d up to which a camera can reliably perform the face detection algorithm (in horizontal direction), using the following relationship [54]:

$$\tan\frac{\alpha}{2} = \frac{a/2}{d} = \frac{(w_{px} * l_{res})/2}{d} \qquad (4.1)$$

Thus, the maximum distance d is 7.9 meters in this specific case (only considering the horizontal view).

64 CHAPTER 4. SYSTEM ARCHITECTURE FOR ACTIVE CAMERAS

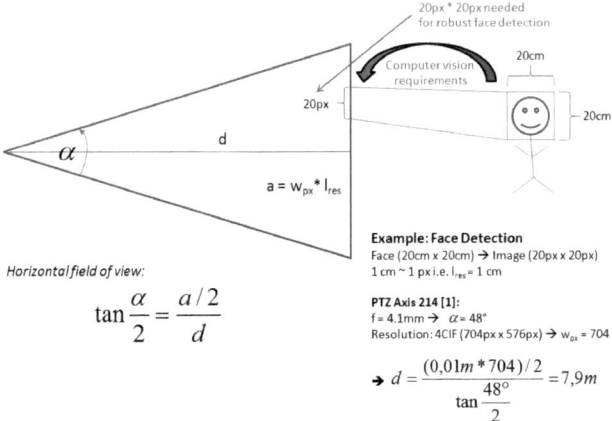

Figure 4.2: Optimal target-to-camera distance

Thus, we are able to derive the requirements for the *Positioning Layer* from the sensing objective including the computer vision algorithm. Figure 4.3, for example, presents the combination of both requirements, i.e. a maximum distance of 7.9 meters and a view angle of less than 15 degrees.

The *Mapping Engine* must also consider the Active Camera's dynamic motion capabilities, e.g. the maximum velocity v_{max}, acceleration a, and the time span until the specific target leaves the camera's actuation range.

Based on this model, the *Mapping Engine* determines the possible points in its actuation range where the Active Camera can travel to before the target leaves. This is triggered each time a target enters the Active Camera's actuation range. After the *Mapping Engine* has determined the possible points for traveling, the *Pose Manager* evaluates them in terms of the imaging quality function, see Figure 4.3, defined by the *User* through the system goals, see Figure 4.1. This is depicted in Figure 4.4 through a heat map. For instance, in case a biometrical algorithm, such as a frontal face detection algorithm, is used, images of targets should be captured at a maximum distance of $7.9\,m$ with a view angle of 0°. Yellow areas denote preferred areas, where the imaging quality will be maximal and the reconfiguration cost minimal. The target's

4.2. ADAPTIVE LOCATION MANAGEMENT ARCHITECTURE

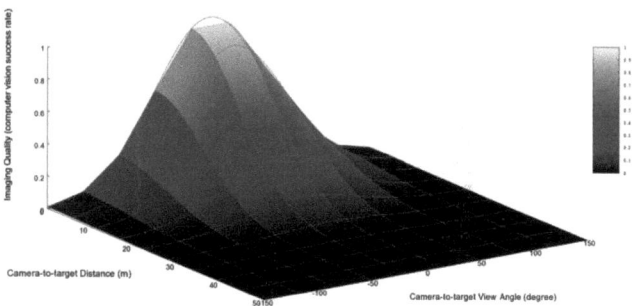

Figure 4.3: An example of the location-dependent quality for observing a target in the camera's actuation range

future positions are extrapolated based on its previous positions. Therefore, the heat map has to be recalculated for every state change of the target. The creation of this heat map is driven by the system goals which are defined by the *User*, e.g. by the aforementioned imaging quality functions. In addition to static parameters, such as the target-to-camera distance or view angle (see Section 5.2.1 for quality functions concerning the *DRofACN* method), dynamic parameters could be added. This can be the current illumination within the actuation range or dynamic obstacles. Nevertheless, this has not been investigated in terms of this thesis.

4.2.5 Layer 3: Coordination

The *Coordination Layer* contains the *Reconfiguration Manager*, which can be configured by the *User* through system goals, see Figure 4.1. In this thesis, the *Reconfiguration Manager* was defined with the system goal to periodically synchronize the Active Camera's clock to avoid inaccuracies in terms of data fusion. In case no infrastructure for clock synchronization is available, the *Reconfiguration Manager* shall chose our reconfiguration method *ACFSync* presented in Section 5.3, which is based on visual events. These visual events can either stem from a visual beacon or from natural events in the surveil-

66 CHAPTER 4. SYSTEM ARCHITECTURE FOR ACTIVE CAMERAS

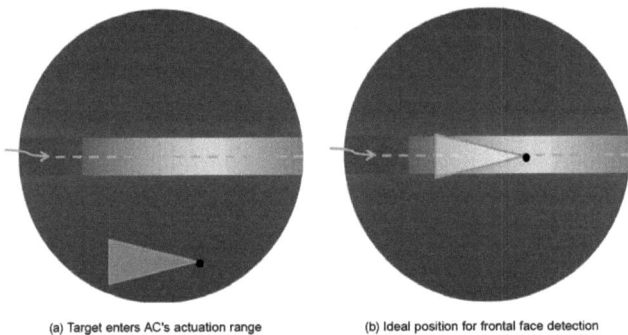

(a) Target enters AC's actuation range   (b) Ideal position for frontal face detection

Figure 4.4: An example of the location-dependent quality for observing a target in the camera's actuation range on the basis of the target's current position

lance area. In the second case, the synchronization has to be performed in a cooperative manner by neighboring Active Cameras.

The second system goal given by the *User* is to capture exactly one image of each target of interest in the surveillance area. This is achieved by the reconfiguration method *DRofACN* which has to be executed by the *Reconfiguration Manager* constantly. It enables the Active Cameras to cooperatively observe multiple, dynamic targets in a cooperative manner. If a target leaves the actuation range of an Active Camera and has not been observed yet, neighboring Active Cameras take over the target to achieve a system-wide observation. The *DRofACN* method is presented in detail in Section 5.2.

The *Target Cache* is used to receive incoming target requests at runtime. As explained in Chapter 3, perceivers communicate with the Active Cameras on a separate communication channel in contrast to the camera-to-camera communication. Incoming target-information-messages (TIM) as depicted in Figure 4.5, e.g. target requests from perceiver nodes or target updates from neighboring cameras, are handled by a separate thread running on each Active Camera in order to add new target requests to the target cache or update existing ones. The TIM messages include spatio-temporal information about the target (e.g. its position and speed, a time stamp and status parameters such as

4.2. ADAPTIVE LOCATION MANAGEMENT ARCHITECTURE 67

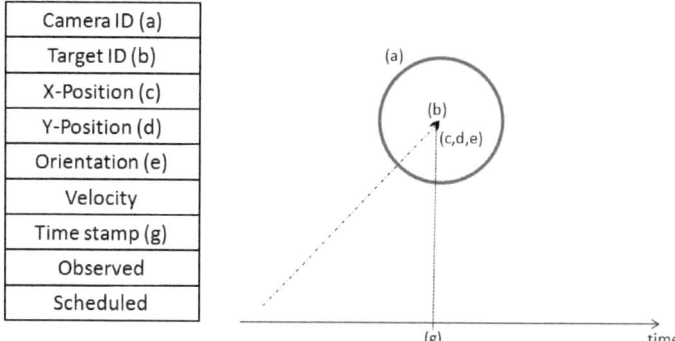

Figure 4.5: Target Information Message (TIM) and corresponding information of a target. The target's ID is unique and defined by the perceiver node detecting the target's first occurrence.

its observation state or if it has been scheduled for observation). The exchange of TIM messages allows the cameras to collaborate on target acquisition, e.g. to avoid that a target is scheduled for observation by two (neighboring) cameras. TIM messages are rather small having a length of 30 Bytes each, since seven fields contain integer values and the last two ones booleans.

4.2.6 Cross-Layer Event Handler

Visual data or, in general, sensor data acquired by the sensors of an Active Camera can be used locally or in a cooperative manner. Therefore, the task of the *Cross-Layer Event Handler* is to allow components of all layers to subscribe to sensorial data. E.g. in [56], the *Mapping Engine* contains an optical compass and is subscribed to the image sensor in order to detect rotation changes by visual means, which cannot be detected by GPS. In addition, sophisticated activity recognition algorithms have been implemented using this concept, see [57]. Thus, components can subscribe to specific activities such as left or right turnings. Furthermore, visual context data acquired by Active Cameras in a cooperative manner could be used for other tasks like key generation for encryption as filed for a patent [59]. Thus, smart activity-based applications beyond the scope of classical camera networks become feasible.

4.3 Summary

This chapter presented a system architecture for Active Camera Networks. Given the requirements for Active Camera Networks, we concluded that a different kind of middleware is necessary, since middlewares for wireless sensor networks are not intended to handle advanced image-processing tasks and sending large amounts of data. On the other hand, adapting a general-purpose middleware is feasible and could fulfill the aforementioned requirements, although the introduced overhead does not yield efficient resource utilization. Therefore, we introduced a new software architecture for Active Camera Networks, which has been implemented as middleware for the ARM chipset, paving the way for scalability and adaptivity in these networks.

The system architecture is encapsulated into four layers and runs on each camera independently. Its distributed *Observer/Controller* design allows for scalability concerning the number of cameras and adaptivity in terms of the environment. *Layer 0* contains the basic functionalities for active sensing and processing such as the capturing and low-level analysis of image data and the physical reconfiguration of the camera's pose. *Layer 1* is responsible for the communication. In *Layer 2*, components are included aiming at determining the optimal pose of an Active Camera for visual surveillance of a corresponding target within the workspace. The objective of *Layer 3* is to handle incoming requests and organize the execution of available reconfiguration methods in order to maximize the system's overall performance.

Based on this architecture, dynamic reconfiguration methods have been developed and integrated in *Layer 3* which are introduced in the following chapter.

Chapter 5

Dynamic Reconfiguration Methods

This chapter presents dynamic reconfiguration methods that make way for self-organization and self-configuration in Active Camera Networks. Since the underlying problem of optimal wide-area target observation in dynamic environments is NP-complete [60], a heuristic approach has been pursued. A method that is approximating solutions to chose the next target to be observed in the most efficient way is *DRofACN*. Afterwards, the reconfiguration method *ACFSync* is presented for active frame synchronization in Active Camera Networks. This method is able to synchronize the frames of neighboring cameras on the basis of visual events. Utilizing visual events for frame synchronization comes along with several advantages compared to traditional synchronization protocols like NTP [50].

In Section 5.1, we will state the formal problem statement for wide-area target acquisition, which is NP-complete. Afterwards, we present the heuristic *DRofACN* in Section 5.2, which approximates solutions for this problem. In Section 5.3, we introduce a method for active frame synchronization called *ACFSync*. Finally, we close with a summary.

5.1 Problem Statement: Wide-Area Target Acquisition

Dynamic reconfiguration for wide-area target acquisition is needed when the number of observation targets exceeds the number of Active Cameras in the surveillance area. In this case, camera scheduling and control become nontrivial, since the following requirements have to be met for visual surveillance:

- none of the given targets should be neglected,
- targets should not receive excessive preference over other targets,
- the quality of captures should be optimized, and
- the assignment of cameras to targets should consider their capacity restrictions (e.g. available observation time) to meet the quality requirements.

Targets may be variable in terms of their speed and entrance points of the Active Camera's actuation range. Thus, the camera has to cope with variable reaction times. In case of multiple targets entering its actuation range at the same time, the reconfiguration problem can be modeled as the classical *Knapsack problem* [61], where the camera's reconfiguration time (including time for repositioning the camera and image processing) represents the knapsack size and the quality of captures the values. These values are defined by the requirements of the underlying image processing function. In case of performing biometric tasks, for example, imagery captured below a maximum distance receives a higher value than imagery captured above this distance. Thus, the goal is that the sum of all values (of those captures) is as large as possible and that there is a capture of every target before leaving the surveillance area. Therefore, this problem is derived from the problem faced by someone who is constrained by a fixed-size knapsack and must fill it with the most useful items.

This means that neighboring cameras should be able to communicate with each other and exchange information about targets. Targets, which have

5.1. PROBLEM STATEMENT: WIDE-AREA TARGET ACQUISITION 71

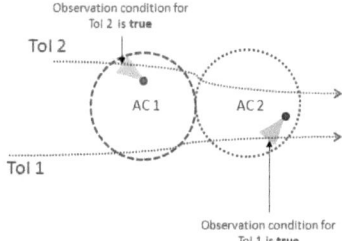

Figure 5.1: Two targets of interest (ToIs) are observed (observation condition is true) in an Active Camera Network of two Active Cameras (ACs) in $[t', t'')$

been observed and detected successfully, should receive lower preference in the neighboring actuation range. This makes way for object observation in wide-area surveillance scenarios such as airports or train stations. Typically, this kind of scenarios consists of tens to hundreds of Active Cameras [62]. Thus, cooperative target acquisition is important in high-load, multi-target scenarios in which one camera is not able to detect all targets in its actuation range due to constrained capacities.

The following section states this problem formally.

5.1.1 Formal Description

In this section, we formulate the problem of dynamic reconfiguration of Active Cameras (ACs) for wide-area target acquisition. The goal is to make exactly one capture of each target of interest (ToI) while moving through the Active Camera Network's surveillance area exceeding a minimum of required imaging quality. Before we state the problem formally, we give an illustrative example: Consider an Active Camera Network of two Active Cameras as depicted in Figure 5.1. In a time interval $[t', t'')$, there are two ToIs assumed to be persons entering the Active Camera Network's surveillance area. ToI 2 is processed by AC 1. Since ToI 1 and ToI 2 enter the actuation range of AC 1 at nearly the same point of time, AC 1 has to decide which ToI to observe. Since it moves to ToI 2 (due to achieving a higher imaging quality than for the other target), ToI 1 cannot be observed and has to be covered by AC 2. Since AC 1 informs AC 2 that ToI 2 has been observed, AC 2 can focus on the observation of ToI 1. Thereby, both ToIs are captured by collaboration before leaving.

The main goal of our reconfiguration method is to optimize the *Target Acquisition Ratio (TAR)* (see Equation 5.1), measuring the number of successfully observed ToIs leaving in relation to the total number of targets, which have entered the surveillance area. In case of $TAR = 1$, there is a high-quality close-up view of each ToI that has entered the Active Camera Network's workspace. The observation condition in Equation 5.2 is evaluated to 1 (i.e. target$_i$.$observed(t) = 1$), if the target has been observed in the time interval $[t', t)$ by an AC j when leaving at time t. This holds, if there is one capture of an AC j, i.e. target$_i$.$observedBy(\text{cam}_j) = 1$, exceeding the minimum required imaging quality q_{min} for $t' \leq t < t''$. Whether the capture of AC j of target i exceeds the threshold q_{min}, is evaluated by the imaging quality function Q_{ij}^{img}. Q_{ij}^{img} is a time-dependent function evaluating the ToI's position while passing the AC j's actuation range. For this purpose, the ToI's position in terms of the view angle and viewing distance is investigated, since both parameters are crucial for successful image processing, e.g. face detection.

Part 2 of the observation condition ($\sum_{\text{cam}_j \in AC}$ target$_i$.$observedBy(\text{cam}_j) = 1$), i.e. assuming that each target is observed exactly *once* before leaving the surveillance area, comes with the following advantages from a practical and theoretical perspective:

Practical issues: The condition that every target has to be observed exactly once allows for local and network-wide load balancing within the camera network. By cooperating on target acquisition, overload situations can be avoided/handled more efficiently, since the exchange of information paves the way for target prioritization. In addition, restricting the number of captures to one makes way for efficient resource utilization, since mechanical problems due to dynamic fatigue of the camera mechanics (e.g. for repositioning) are reduced.

Theoretical issues: Furthermore, the observation condition is used to

5.1. PROBLEM STATEMENT: WIDE-AREA TARGET ACQUISITION

prove the NP-hardness in Section 5.1.2. The problem of wide-area target acquisition is equivalent to the Hamiltonian path problem, since an instance of the Hamiltonian path problem is given by a graph $G = (V, E)$ and a vertex v, and it has to be decided if there exists a path starting from v that visits all vertices exactly once. This is similar to an instance of the wide-area target acquisition problem, since an AC could be placed at a vertex v and the remaining $|V| - 1$ vertices could be the designated targets. Thus, an edge means that the target is observed with an imaging quality exceeding q_{min} and before leaving the surveillance area.

Formally, the problem stated above can be defined as follows:

Given n ToIs (targets of interest) entering and leaving the surveillance area of an Active Camera Network in the time interval $[t', t'']$ (time is assumed to be discrete and represented as non-negative integers):

Maximize:

$$TAR = \frac{1}{n} \sum_{t=t'}^{t''} \sum_{\text{target}_i \in \text{targets}} \text{target}_i.observed(t) \qquad (5.1)$$

Subject to:

$$\text{Observation condition} \qquad (5.2)$$

(1) Observed before leaving the surveillance area:

$$\text{target}_i.observed(t) = 1 \leftrightarrow \exists \text{ cam}_j \in AC :$$
$$Q_{ij}^{img} > q_{min} \wedge \text{target}_i.observedBy(\text{cam}_j) = 1$$

(2) Observed by exactly one AC:

$$\forall \text{target}_i :$$
$$\sum_{\text{cam}_j \in AC} \text{target}_i.observedBy(\text{cam}_j) = 1$$

74　　CHAPTER 5. DYNAMIC RECONFIGURATION METHODS

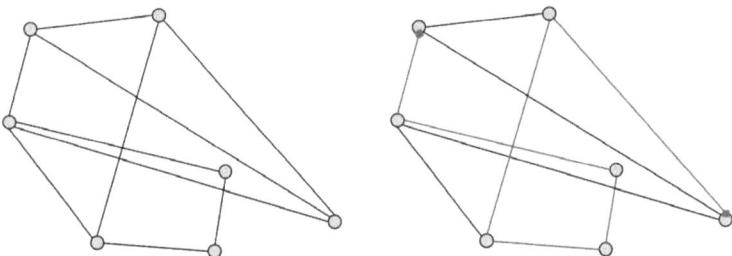

Figure 5.2: A Hamiltonian path (red) over a graph.

5.1.2 Proof of Problem Complexity

Optimizing the target acquisition ratio (TAR) for wide-area target acquisition under the observation condition is NP-hard, as shown by the following theorem.

Theorem 1: There is no polynomial time algorithm for solving the wide-area target acquisition problem optimally with the TAR objective under the observation condition, unless P = NP.

Proof: We show that a polynomial time algorithm for wide-area target acquisition with the TAR objective implies a polynomial time algorithm for Hamiltonian path, a well known NP-complete problem.

In the mathematical field of graph theory, a Hamiltonian path (or traceable path) is a path in an undirected graph that visits each vertex exactly once, see Figure 5.2. A Hamiltonian cycle (or Hamiltonian circuit) is a cycle in an undirected graph which visits each vertex exactly once and also returns to the starting vertex. There is a simple relation between the two problems. The Hamiltonian path problem for graph G is equivalent to the Hamiltonian cycle problem in a graph H obtained from G by adding a new vertex and connecting it to all vertices of G. The Hamiltonian cycle problem is a special case of the traveling salesman problem, obtained by setting the distance between two cities to a finite constant if they are adjacent and infinity otherwise.

Our proof is analog to the one used by Lagoudakis et al. [63] who pre-

5.1. PROBLEM STATEMENT: WIDE-AREA TARGET ACQUISITION

sented that the multi-robot routing problem is NP-complete. Without loss of generality, we make the following assumptions for simplification, which do not lead to an overall reduction of the problem complexity:

1. The Active Camera Network consists of one AC only and its actuation range $A_r = \infty$ (wide-area).

2. Time is not considered.

Considering time or adding more ACs with restricted actuation ranges increases the problem complexity, since the combinatorial complexity rises, e.g. by adding time the correct sequence of observations has to be found or by adding cameras the allocation of ToIs to cameras has to be determined. Thereby, the problem is transferred into the knapsack problem as explained in Section 5.1. Nevertheless, this is not considered in this proof.

An instance of a Hamiltonian path problem is given by a graph $G = (V, E)$ and a vertex v, and it has to be decided if there exists a path starting from v that visits all vertices exactly once. We reduce it to an instance of wide-area target acquisition as follows. Let $G' = (V, c)$ be the complete (i.e. contains an edge between all pairs of vertices) weighted graph on V with costs $c(u, w) = 1$, if $(u, w) \in E$ of the graph G, and $c(u, w) > 1$, otherwise. The AC is placed at vertex v and the remaining $|V|-1$ vertices are designated as ToIs. The weights in G' satisfy the triangle inequality and a weight of one, $c(u, w) = 1$, means that the ToI is observed before leaving the surveillance area with an imaging quality exceeding q_{min} (observation condition (1)). A Hamiltonian path in G exists if and only if the optimal solution for wide-area target acquisition in G' has a cost of $|V| - 1$, i.e. every ToI is visited only once (observation condition (2)). In case a ToI is visited twice, the path is not a Hamiltonian path. A Hamiltonian path in G is also an optimal solution for wide-area target acquisition in G' with cost $|V| - 1$, since this means the observation condition is fulfilled for each ToI and $TAR = 1$. Vice versa, if G does not have a Hamiltonian path, then any path in G' that starts from v and visits all the vertices exactly once has to use some edge of cost greater than one in G' or of one edge twice. Therefore, the cost of an optimal solution will be at least $|V|$.

∎

CHAPTER 5. DYNAMIC RECONFIGURATION METHODS

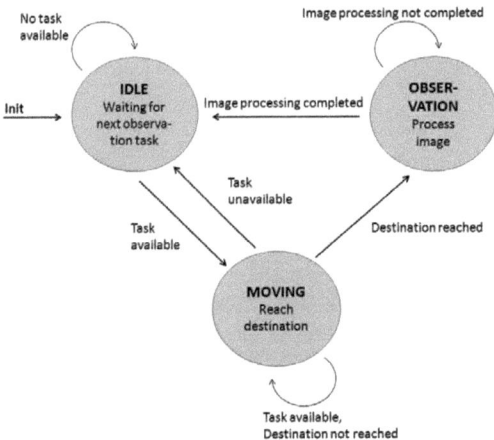

Figure 5.3: State machine of the *DRofACN* method

Given this hardness result, the following section introduces a lightweight heuristic for dynamic and distributed reconfiguration for solving instances of the wide-area target acquisition problem.

5.2 DRofACN

DRofACN is a distributed control algorithm for dynamic reconfiguration of cooperating Active Cameras (ACs) [64]. The basic idea of *DRofACN* is that a set of ACs collaborates for acquiring close-up views of targets in a surveillance area. The AC control is based on the output of perceiver nodes, i.e. generated target requests, estimating the position of salient targets, i.e. ToIs (see Section 3.2), within the ACs' actuation ranges. The main goal is to acquire views for object tracking or for biometric purposes, such as face detection. E.g. in case of face detection, there should be one capture of each ToI's face before it leaves the Active Camera Network's surveillance area. In case of object tracking, targets have to be captured in a way alleviating the process of foreground-

5.2. DROFACN

background subtraction for trajectory composition.

DRofACN is based on a state-machine which will be described in detail in the following. Figure 5.3 shows an overview about the different states of the state-machine. A timer is initialized (see Algorithm 2 line 3) and the AC is set to *IDLE* mode waiting for observation tasks. In case the timer expires (Algorithm 3 line 1), all target requests within the target cache are processed in order to find the next observation task, i.e. the most-promising target request for image acquisition. A timer is chosen in order to reduce CPU load by triggering the scheduling process in discrete time steps only. Thus, it allows for scalability in overloaded scenarios. In case a target request is found exceeding a pre-defined imaging quality threshold, the AC changes to *MOVING* mode. In this mode, the AC informs neighboring ACs about its intention to observe the target and starts to move towards the target in order to acquire close-up views. After arriving, the AC enters *OBSERVATION* mode in order to acquire high-quality imagery. Afterwards, it returns to *IDLE* mode.

5.2.1 Asynchronous Scheduling Process

In this section, we present the scheduling process, which is performed asynchronously to allow for short reaction times in dynamic environments. The scheduling process operates on data in the target cache stemming from the perceiver nodes. These communicate with the Active Cameras on a separate communication channel in contrast to the camera-to-camera communication. Incoming target-information-messages (TIM), as explained in Section 4.2.5, are handled by a separate thread running on each Active Camera, see Algorithm 2.

The TIM messages include spatio-temporal information about the target (e.g. its position and speed, a time stamp and status parameters such as its observation state or if it has been scheduled for observation). The exchange of TIM messages allows the cameras to collaborate on target acquisition, e.g. to avoid that a target is scheduled for observation by two ACs. Incoming messages are processed immediately on their arrival. Thus, part 2 of the observation

---
Algorithm 2 DRofACN - Target Cache Thread

1: **init:**
2:   *set ID //unique device identifier of local AC*
3:   *set timer //triggering replanning each* $500ms$
4:   *init TC //start with empty target cache*
5:   *init A_r, p //p with orientation (x, y, z, α)*
6:   *init state $\leftarrow IDLE$*
7:   *$tr_{cur} \leftarrow$ empty, $q_{cur} \leftarrow 0$*
8:
9: **on** *incoming TIM : //from perceivers/ACs*
10:   **if** *TIM.targetID $\in TC$*
11:     *update TC //data from neighbors*
12:   **else** *//new target*
13:     *add targetRequest to TC*
14:   **end if**
15:
16:   *//conflict with neighboring AC*
17:   **if** *$tr_{cur}.targetID \equiv TIM.targetID \wedge TIM.scheduled$*
18:     *$tr_{cur}.cameraID \leftarrow \min(ID, TIM.cameraID)$*
19:     **if** *$tr_{cur}.cameraID \neq ID$ //neighbor wins*
20:       *state $\leftarrow IDLE$ //stop moving*
21:       *$q_{cur} \leftarrow 0, \; tr_{cur} \leftarrow$ empty //start replanning*
22:     **else**
23:       *//nothing to do*
24:     **end if**
25:   **end if**
26: **end on**

5.2. DROFACN

condition as defined in Equation 5.2 is fulfilled. Nevertheless, conflicts can occur in case a neighboring Active Camera has scheduled the same ToI for observation, see Algorithm 2 line 16. In this case, the conflict is resolved as follows: The ToI is allocated to that camera possessing the lower device identifier, see Algorithm 2 line 18. The other camera starts replanning.

Algorithm 3 DRofACN - Asynchronous Scheduling Process

1: **on** $timer expire$:
2:    $q_{max} \leftarrow 0$, $tr_{max} \leftarrow empty$
3:    **forall** $tr_i \in TC$ **do**
4:      **if** $tr_i.observed \equiv true \wedge tr_i$ is about to leave A_r
5:        broadcast $TIM(tr_i)$ //$forward\ observed\ targets$
6:        delete $TC.tr_i$
7:      **else** //$examined\ w.r.t.\ imaging\ quality$
8:        $q_i \leftarrow \exp^{c_1 \cdot d^2} * \neg tr_i.scheduled$ //$d: target\ distance$
9:        **if** $q_i > q_{cur}$
10:          **forall** $p_j \in A_r$ **do**
11:            $t_p \leftarrow \text{now}() + \text{predictArrivalOfCam}(p_j)$
12:            $p_{tr} \leftarrow \text{predictTargetPos}(tr_i, t_p)$ //$with\ orientation$
13:            $q_{temp} \leftarrow Q^{img}(p, p_{tr})$
14:            **if** $p_{tr} \in A_r \wedge q_{temp} > q_{max}$
15:                $q_{max} \leftarrow q_{temp}$, $tr_{max} \leftarrow tr_i$
16:            **end if**
17:          **end forall**
18:        **end if**
19:      **end if**
20:    **end forall**
21:    **if** $q_{max} > q_{cur} \wedge q_{max} > q_{min} \wedge \neg tr_{max}.scheduled$
22:      $q_{cur} \leftarrow q_{max}$, $tr_{cur} \leftarrow tr_{max}$
23:      $tr_{cur}.scheduled \leftarrow true$ //$is\ scheduled$
24:      broadcast $TIM(tr_{cur})$ //$inform\ all\ neighbors$
25:      $state \leftarrow MOVING$
26:    **end if**
27:    set $timer$ //$reset\ timer$
28: **end on**

Based on the data in the target cache, the scheduling process is executed each time the timer expires, see Algorithm 3 line 1. This *timer* is set after initialization of the AC, see Algorithm 2 line 3. The scheduling process is responsible for finding the ToI with the highest imaging quality. In case a ToI exists in the target cache that has already been observed, i.e. by a neighboring AC, and the ToI is about to leave the AC's actuation range, see Algorithm 3 line 4, the target request is broadcasted to all neighboring ACs (local neighbors only). Thus, knowledge about observed targets is transmitted to the local neighbors. Afterwards, it is deleted from the AC's target cache.

If the ToI has not been observed yet, it is examined in terms of its imaging quality, see Algorithm 3 line 7. For this purpose, the target-to-camera distance is computed and evaluated according to the function depicted in Figure 5.4 ($f(d) = \exp^{c_1 \cdot d^2}$). Generally, the quality of images captured by an AC degrades as the distance d between the target and AC increases. The parameter c_1 is chosen empirically and depends on the AC's pre-defined actuation range. The second factor, in Algorithm 3 line 8, considers if the ToI has already been scheduled by a neighboring AC. If it has been scheduled, this factor evaluates to zero and thus the imaging quality, too. This mechanism avoids that ToIs are scheduled twice. In case a neighboring AC has scheduled the ToI but this information has not been disseminated through the network, there are two mechanisms to resolve this conflict: (1) After selecting the ToI for observation, a final check based on the updated target cache data is performed whether the ToI has not been scheduled by another AC, see Algorithm 3 line 21, and (2) in case the ToI has been scheduled for observation, the conflict is resolved when the *TIM* of the neighboring AC arrives, see Algorithm 2 line 16.

If the distance of the ToI stored in tr_i is lower than all previous ones in the target cache, the arrival times for all points in its actuation range are computed, Algorithm 3 line 11, and the target position is estimated for these times, see Algorithm 3 line 12. The target-to-camera distance and the view angle are computed on the basis of all future positions of the target. These positions are predicted based on linear extrapolation, i.e. the most recent two target requests are utilized for calculating the moving direction and velocity of the target. Based on this data, a linear extrapolation is performed. In addition, this extrapolation could also be based on more sophisticated data such as aggregated trajectories, see [65]. Nevertheless, this is beyond the scope of this thesis. The imaging quality can be estimated according to the following

5.2. DROFACN

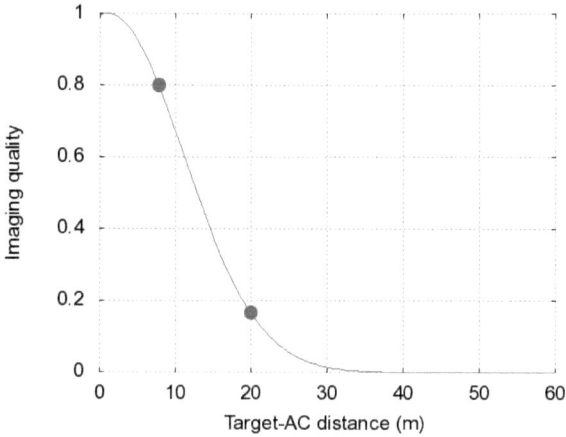

Figure 5.4: Quality function of target-to-camera distance [6] (red points correspond to computer vision success rates for the x-values respectively and are supporting points for the construction of the quality function)

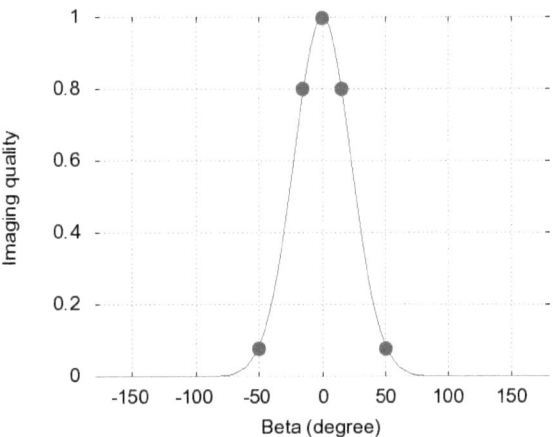

Figure 5.5: Quality function of the view angle [7] (red points correspond to computer vision success rates for the x-values respectively and are supporting points for the construction of the quality function)

function:

$$Q^{\text{img}} = q^{\text{dist}} * q^{\text{angle}} \qquad (5.3)$$

A specific imaging quality function can be as follows:

$$Q^{\text{img}} = \exp^{c_1 \cdot d^2} * \exp^{c_2 \cdot \beta^2} \qquad (5.4)$$

whereas d is computed and evaluated as described before and the view angle $\beta = tr_i.p(\alpha) - p(\alpha)$ is evaluated according to the function depicted in Figure 5.5 (α is the orientation of the target and AC respectively).

If there exists a point within the AC's actuation range exceeding the imaging qualities of all points of previous target requests, it is stored in tr_{max}, see Algorithm 3 line 15. After evaluating all target requests and observation points for each target respectively, a final check on the updated target cache

5.2. DROFACN

is performed, see Algorithm 3 line 21, and the target request with the highest imaging quality is scheduled for observation, see Algorithm 3 line 23. This is disseminated to all neighboring ACs (local neighbors only) and the AC enters the mode *MOVING*. During being in this mode, the AC can still be interrupted to resolve conflicts with neighboring ACs.

An important property of *DRofACN* is that it does not require a closed-form expression for the imaging quality function, but may proceed using only a numerical computation. Thus, *DRofACN* can be utilized in realistic scenarios, where nonlinearities can arise in terms of the imaging quality function, e.g. due to static or dynamic obstacles. Nevertheless, this is not considered in this thesis.

5.2.2 IDLE Mode

Algorithm 4 IDLE Mode
1: wait for next observation task

In *IDLE* mode, the AC waits for the next observation task. *IDLE* mode is joined initially and after successful image processing from *OBSERVATION* mode. If the AC is in *MOVING* mode and the task it is moving to becomes obsolete, e.g. because it is scheduled by a neighboring AC, it changes to *IDLE* mode, too.

5.2.3 MOVING Mode

Algorithm 5 MOVING Mode
1: **while** $tr_{cur}.cameraID \equiv ID$ **do**
2:    change own position to tr_{cur}
3: **end while**
4:
5: **on** $targetDestinationReached$:
6:    $state \leftarrow OBSERVATION$
7: **end on**

The *MOVING* mode is entered from *IDLE* mode. An AC turns to *MOVING* mode in case an observation task, i.e. a ToI stored in tr_{cur}, is chosen for observation. When the *MOVING* mode is entered, the AC begins moving toward the calculated best position for capturing close-up views of the ToI. In case the target is reached, i.e. the event *targetDestinationReached* occurs, the AC changes its state to *OBSERVATION*. While the AC is in state *MOVING*, it continues with the scheduling process as described in the previous section. Thereby, the AC is highly adaptive in terms of changes in its environment, since it can revise its decision at any time in case a ToI with a higher observation quality appears. A rescheduling is only performed if the computed imaging quality of a new target is greater than the current one, i.e. the distance of the new target has to be lower than the one of the current target. Nevertheless, the more the AC approaches the ToI the more unlikely this is, since the quality of the ToI increases. In addition, by choosing the greater operator, possible oscillations in terms of rescheduling are avoided.

5.2.4 OBSERVATION Mode

Algorithm 6 OBSERVATION Mode

1: **while** $tr_{cur}.cameraID \equiv ID$ **do**
2:     capture imagery of ToI
3: **end while**
4:
5: **on** $imageProcessingFinished$:
6:     $tr_{cur}.observed \leftarrow true$
7:     broadcast $TIM(tr_{cur})$ //to all neighbors
8:     $q_{cur} \leftarrow 0$, $tr_{cur} \leftarrow empty$
9:     delete $TC.tr_i$
10:     $state \leftarrow IDLE$
11: **end on**

An AC enters the *OBSERVATION* mode, if it has reached its predetermined position for observation. Then, it starts capturing high-quality imagery in order to process them. As described before, the quality of the imagery depends on the view angle and distance. After finishing the image processing, e.g. a face detection algorithm, the observed condition of the target request for the corresponding ToI is set to *true* and an update is broadcasted to all neighboring ACs (local neighbors only). Thereby, they can update their target cache and avoid the redundant observation of ToIs. Afterwards, the current observation task is set to *empty* and the AC enters the $IDLE$ mode in order to wait for the next observation task.

5.2.5 Correctness

In this section, we show that *DRofACN* allows the target detection ratio (TAR) to increase monotonically due to fulfilling the observation condition. Even though ACs may plan their observation tasks without any global coordination, the observation condition is not violated. Additionally, TAR will reach a local maximum, though not necessarily the global one, which is hard to ensure without global coordination. We prove *DRofACN*'s capability to fulfill the *observation condition* at all times.

Theorem 2 (Correctness of DRofACN): Under *DRofACN* execution, the observation condition is always fulfilled, i.e. each target is observed at most once with a sufficient imaging quality.

Proof: The validity of this theorem can be shown with the help of the Algorithms in Section 5.2.1-5.2.4. While moving through the surveillance area, a target traverses the actuation ranges (A_r) of different ACs. Since the actuation ranges cover the whole surveillance region and may overlap themselves, the target is at least inside one A_r at a time. This proof distinguishes between the target residing in only one AC's actuation range (1) and residing in more than one AC's actuation ranges (2).

1. If a target becomes salient (see Section 3.2) while being in exact one A_r only or enters only one A_r while being salient, a notion of this target exists only in one AC's target cache. Depending on the current situation, this AC may schedule the target.

    (a) If it *does* schedule the target, the AC sends a TIM. Thus, the neighbors know about the target being scheduled/captured. This ensures the observation condition.

(b) Otherwise, the target leaves A_r without being scheduled/observed. This means, it enters at least one actuation range of a neighboring AC and case (1) or (2) reapply for the target depending on the actuation range's topology.

2. If a target becomes salient while moving through more than one A_r or enters more than one A_r while being salient, all concerned ACs may schedule the target during its presence. Without loss of generality, the ACs are numbered in the order in which they enter their *on timerexpire* methods whose run schedules the target in question.

(a) If the first camera enters its *on timerexpire* method more than δ_{lat} (worst-case latency for sending a TIM message, see Section 3.1) before all other ACs enter it, AC 1 schedules the target and sends the TIM. Because of the message delay of δ_{lat}, all other ACs, which may schedule the target, too, receive the message before entering their *on timerexpire* method. These can only be neighboring nodes due to our communication model. Thus, they update their target cache appropriately and cannot schedule the target anymore. Thereby, the observation condition holds.

(b) A set of ACs entering their *on timerexpire* methods in a time interval $[t', t'']$ with $t'' - t' < \delta_{\text{lat}}$, which is shorter than the message propagation time, may lead to temporary inconsistent target caches. This happens, because all of them may schedule the target, see Algorithm 3 line 21, if no better target is in range. All ACs, which scheduled the target, send a TIM message and change to *MOVING* mode. While moving, each AC receives all of the TIM messages (after the message propagation delay) and updates its target cache accordingly. Lines 16 to 25 in Algorithm 2 ensure a clean conflict resolution, since only the camera with the smallest identifier remains to have the target scheduled. All other cameras return to the *IDLE* state and plan their next target. They cannot reschedule the target, since their target caches regained consistency w.r.t. to the target, and the evaluation of the target's quality in Algorithm 3 line 8 will return zero. The worst case w.r.t. the consistency condition is one or more ACs, which schedule the target concurrently and are located in the optimal position to capture an imagery of the target. This would lead to an immediate change from the *IDLE* mode to the *OBSERVATION* mode. However, the ACs will regain target cache consistency upon TIM delivery. Since observation time $\gg \delta_{\text{lat}}$ (message delay assuming a wireless sensor network for data exchange), no AC could have finished the capturing. Therefore, the conflict resolution (see above) leads to only one remaining AC capturing the target. Thus, the observation condition holds. ■

5.2.6 Phenomena Adaptivity (ENRA)

In this section, we present a mechanism called *ENRA* for paving the way for phenomena adaptivity of the aforementioned *DRofACN* algorithm [66]. Based on the number of events occurring in the workspace, ACs increase or decrease so-called *spatial redundancy regions* with neighboring nodes to balance the network's load. As will be shown in Section 6.2.6, this mechanism is able to increase the overall system's performance.

Control of the Center of Movement

For this purpose, we assume that the AC can additionally control its center of movement d in terms of its internal state $AC.state = (d, A_r, p)$, see Section 3.1.2. Thereby, the actuation range is influenced by the AC's actuation radius a_r and its center of movement (circular area around d). $p = (x, y)$ is the AC's current position (x, y). Overlapping actuation ranges of neighboring cameras are called *spatial redundancy regions*, which can be utilized by the cameras to cooperate on improving the system's utility and where conflicts with neighboring cameras, i.e. so-called *spatial conflicts*, can occur. By adapting the center of movement $AC.state.d$ at runtime, ACs are able to create these regions based on self-organization.

ENRA

ENRA's objective is to find the optimal network configuration, i.e. the optimal position of the ACs' center of movement, in order to increase the performance of the *DRofACN* algorithm, which is responsible for choosing the next target for observation.

In order to calculate the optimal center of movement, the number of events is counted which occurred within the actuation range of the node and its neighbors. This is triggered in case a target request arrives (see Algorithm 7 line 9) and relevant data is stored in the matrix E (see Algorithm 7 line 16-20). The weight factors, i.e. 0.1 and 0.9, define whether the node should focus on creating spatial redundancy regions with neighboring nodes or on maximal coverage. In our case, 0.1 and 0.9 are chosen such that higher priority is set on creating spatial redundancy zones. Based on the data in E, the optimal center of movement is computed when the timer, i.e. the time to update (TTU), expires (Algorithm 7 line 23).

When the timer expires, the reconfiguration process is executed. For this purpose, the utility of each point within the node's potential actuation range $A_{r_{pot}}$ is calculated. $A_{r_{pot}}$ contains the points for all possible center of movements, i.e. $AC.state.d \pm \delta \ \forall \ 0 \leq \ \delta \ \leq n_d \cdot a_r$, see Algorithm 7 line 24. n_d is called the *node distance*. $E[p]$ contains all data in range of the AC.

Based on the computed utilities of each point, the optimal center of movement is computed in Algorithm 7 line 27-32. In order to achieve this, the weighted arithmetic mean is computed for each row and column of the matrix U. Afterwards, the mean value of these elements is computed and the new optimal position is achieved as illustrated in Figure 5.6. This new position is used for reconfiguration and sent to all neighboring nodes.

ENRA is a very basic mechanism for phenomena adaptivity in camera networks. Nevertheless, it could be replaced by more sophisticated mechanisms for the analysis of spatio-temporal recurring patterns, e.g. as presented by Sester et al. in [55].

Algorithm 7 ENRA - Reconfiguration Process

1: **init:** //added to the $DRofACN\ configuration$
2: $\quad set\ timer \leftarrow TTU$ //e.g.10s
3: $\quad init\ E \leftarrow empty,\ U \leftarrow empty$
4:
5: **on** $incoming\ COMmsg:$ //from neighboring ACs
6: $\quad$ update neighborhood cache
7: **end on**
8:
9: **on** $incoming\ TIM:$ //from perceivers/ACs
10: $\quad$ **if** $TIM.targetID \in TC$
11: $\quad\quad$ update TC //data from neighbors
12: $\quad$ **else** //new target
13: $\quad\quad$ add targetRequest to TC
14: $\quad$ **end if**
15: $\quad$ //update E
16: $\quad$ **if** $p = TIM.(x,y) \in A_r$ of neighbors
17: $\quad\quad E[p] \leftarrow E[p] + 0.9$ //for spatial redundancy
18: $\quad$ **else**
19: $\quad\quad E[p] \leftarrow E[p] + 0.1$ //for maximal coverage
20: $\quad$ **end if**
21: **end on**
22:
23: **on** $timerexpire:$ //reconfiguration process
24: $\quad$ **forall** $p = (x,y) \in A_{r_{\text{pot}}}$ **do**
25: $\quad\quad U[p] \leftarrow U[p] + E[p], E[p] \leftarrow 0$
26: $\quad$ **end forall**
27: $\quad$ **foreach** $row\ i \in U$ **do**
28: $\quad\quad X[i] =$ compute weighted mean of i
29: $\quad$ **end foreach**
30: $\quad$ **foreach** $column\ j \in U$ **do**
31: $\quad\quad Y[j] =$ compute weighted mean of j
32: $\quad$ **end foreach**
33: $\quad d = (\text{mean } X,\ \text{mean } Y)$
34: $\quad$ change COM to d
35: $\quad$ send COMmsg d to neighbors
36: $\quad set\ timer$ //reset timer to TTU
37: **end on**

90 CHAPTER 5. DYNAMIC RECONFIGURATION METHODS

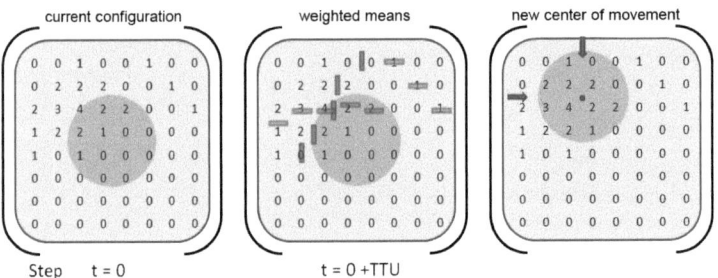

Figure 5.6: Computation of the best center of movement

5.3 Active Frame Synchronization

In this section, we introduce a method called *ACFSync* (Active Frame Synchronization of Active Cameras), which is able to synchronize the frames of Active Cameras (ACs) on the basis of visual events. Utilizing visual events for frame synchronization comes along with several advantages compared to traditional synchronization protocols. Traditional synchronization protocols are based on the exchange of communication messages and the estimation of transmission times. Nevertheless, these methods assume symmetric communication links and suffer from non-deterministic errors like message delays [67]. Wireless communication services, e.g. such as General Packet Radio Service (GPRS) in 3G cellular communication systems, may suffer from high asymmetry in terms of the communication link as depicted in Figure 5.7. 1,000 packets have been sent from a mobile phone (client) using a GPRS link to a server with Internet connection (Desktop PC) and back and the single trip times have been measured based on the initially synchronized clocks of client and server. As illustrated in Figure 5.7, one can see that GPRS possesses a strong link-asymmetry. For strong link-symmetry, the results need to form a line through origin.

ACFSync is only based on image information and provides a solution for Active Camera Networks without suffering from synchronization errors due to message-based communication, since image information propagates with speed

5.3. ACTIVE FRAME SYNCHRONIZATION

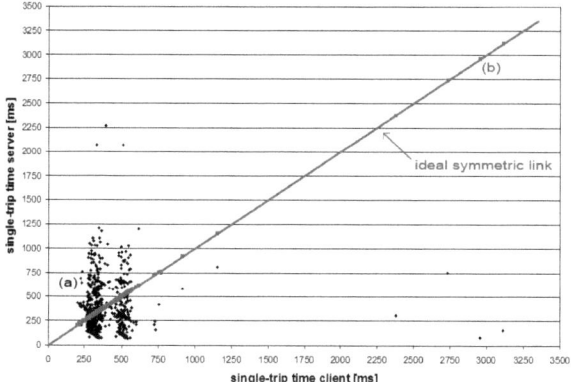

Figure 5.7: Ratio of client and server single trip times: (a) Asymmetric GPRS network link (b) Ideal symmetric link (line through origin).

of light and arrives at the image sensor of cameras sharing the same field of view nearly at the same point of time. Traditional message-based communication is only used to exchange status messages but not for extracting timing information for time synchronization. *ACFSync* utilizes the spatio-temporal properties of the events for estimating the internal frame offset. Visual events can be actively triggered by an optical beacon or passively by moving target objects within the surveillance area.

Physical time plays an important role in Active Camera Networks, since the basic operation is *data fusion*. Data from multiple ACs is agglomerated to form a single meaningful result. In order to fuse captured frames, they need to be timestamped in the first place. Practically, there exists a latency for capturing a frame, e.g. due to exposure time. This time delay is determined by the camera's frame rate. E.g. in case of a frame rate of $25\,fps$, this time delay is $40\,ms$. This means that every frame has to be timestamped with a time interval $[t, t + 1/fps]$ (t stems from the AC's internal clock). In order to fuse data of several ACs, these time intervals have to be synchronized. This process is called *frame synchronization*. In order to avoid time ambiguity during frame synchronization, this can be achieved as follows:

- Synchronizing the ACs' internal clock (i.e. managing the t of the time interval $[t, t + 1/fps]$ by clock synchronization mechanisms)

- Aligning the frames of multiple ACs (i.e. aligning the time intervals $[t, t + 1/fps]$ to each other by data alignment mechanisms)

After presenting the problem statement in Section 5.3.1, we introduce our method in Section 5.3.2. Our method is able to align video sequences of neighboring cameras. In addition, it is able to synchronize the camera's internal clock to an optical beacon signal.

5.3.1 Problem Statement: Frame Synchronization

The problem of frame synchronization is stated in this section formally. We assume to have two video sequences S and S', whereas S is the reference signal the video sequence S' is going to be synchronized to. For frame synchronization, we have to find the temporal transformation for correlating both sequences. For this purpose, we have to solve the following equation whereas $c_S(t)$ is the time of capturing the reference sequence S, $c'_S(t)$ is the time of capturing the sequence S', Δt is a constant time offset and s is the drift:

$$c'_S(t) = \Delta t + s \cdot c_S(t) \tag{5.5}$$

The goal is to find the time offset Δt between the clocks $c_S(t)$ and $c'_S(t)$, which corresponds to the frame offset (measured in frames) of both sequences.

5.3.2 ACFSync: Active Camera Frame Synchronization

In this section, we present the reconfiguration method for visual event-based frame synchronization of Active Cameras - called *ACFSync*. It is based on image information only and provides a solution for Active Camera Networks without suffering from synchronization errors due to message-based communication. Visual events and their spatio-temporal properties are used for synchronization. Thus, the propagation time is removed from the time-critical

5.3. ACTIVE FRAME SYNCHRONIZATION

path, which dominates the delay in wide-area networks (e.g. including the queuing and switching delay at each router as the message transits through the network). This method consists of two distributed algorithms and combines a *sender-to-receiver* and *receiver-to-receiver* approach (see Section 7.4) to allow for scalability and robustness. It enables participating ACs to synchronize their clocks with reference to an optical broadcast signal (sender-to-receiver approach) and reason about frame offsets of neighboring ACs sharing the same field of view (receiver-to-receiver approach).

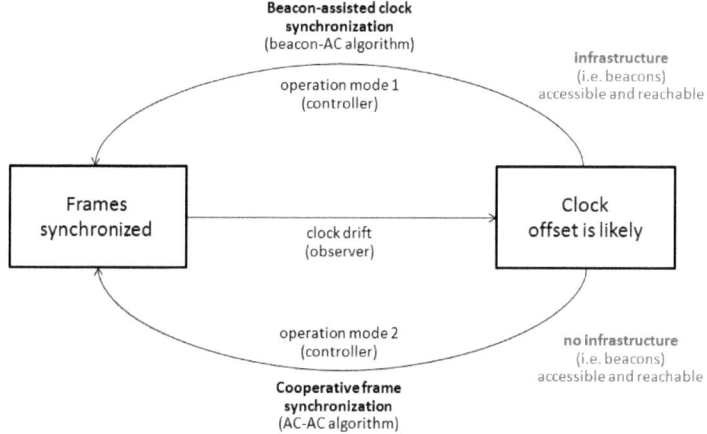

Figure 5.8: Overview about the *ACFSync* method

The basic idea of the algorithm is not to synchronize the local clocks of the ACs through traditional message-based synchronization schemes, but instead use visual cues of the environmental scene. Visual events correspond to communication messages, whereas the visual event is the sender and ACs in visual range are the receivers. The advantage of using visual events is that they propagate with speed of light and consequently arrive at each image sensor at approximately the same point of time (and so reducing the send, access and propagation time from the time-critical path). The method proposed in this thesis consists of two algorithms as depicted in Figure 5.8 allowing for robust, scalable, and accurate frame synchronization in Active Camera Networks:

1. Beacon-assisted clock synchronization algorithm (operation mode 1)

2. Cooperative frame synchronization algorithm (operation mode 2)

Operation mode 1 requires infrastructure in form of optical beacons. Operation mode 2 does not require such an infrastructure, since it is based on visual events which are triggered by moving targets and frames are synchronized in a cooperative manner. By combining both approaches, a robust method for visual event-based frame synchronization of Active Cameras is obtained.

Concerning the beacon-assisted clock synchronization (operation mode 1), visual events are represented by blinking beacons (based on LEDs). The blinking of the beacons is controlled in order to transmit messages, e.g. numerical values, through the optical channel. By sending these messages at specific points of time, temporal information is piggybacked. Thus, the beacons are utilized as senders broadcasting time stamps. Theses time stamps are received by ACs in visual range (receivers) and decoded to obtain the encoded time stamp. Since the optical messages are received at all receivers nearly at the same point of time due to the low propagation time, the point of time of sending this message can be correlated with the time stamp. Afterwards, the local clocks are set. Re-synchronization is needed frequently due to the AC's clock drift. For example, in case of drifting $10\,\mu s$ per second (see Section 3.1.3), the ACs have to be synchronized once an hour in order to avoid mis-assignments of frames of neighboring cameras sampling their environment with 25 frames per second.

Cooperative frame synchronization (operation mode 2) is used, if no beacons are accessible/reachable by the ACs or operation mode 1 is likely to deliver error-prone results due to high visual load in the surveillance area. Operation mode 2 uses visual events for frame synchronization stemming from moving targets, e.g. humans moving through the cameras' field of view. Moving targets generate spatio-temporal information in the cameras' field of view observing the same event. Since visual information propagate with speed of light, it is received at each camera at approximately the same point of time. The

5.3. ACTIVE FRAME SYNCHRONIZATION

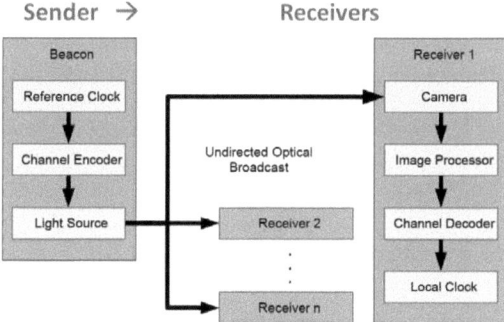

Figure 5.9: Conceptual overview about the sender-to-receiver approach based on beacon-assisted clock synchronization

visual event's spatio-temporal data is extracted by computer vision algorithms and utilized to cooperatively reason about the cameras' clock offset.

The following two subsections present our synchronization methods in detail.

5.3.3 Beacon-assisted Clock Synchronization Algorithm

Figure 5.9 gives an conceptual overview about the sender-to-receiver approach based on the beacon-assisted clock synchronization. The beacon is the sender and contains a *Reference Clock*. If the *Reference Clock* is synchronized to UTC, external synchronization is possible. The *Channel Encoder* transforms the beacon's local time into time stamps consisting of so-called *optical bits*. Optical bits are transferred to rectangular brightness signals by the *Light Source*. A sequence of optical bits forms an optical message containing the time of the beacon at transmission time. This message is sent as undirected broadcast via the optical channel. All receivers, i.e. ACs in visual range, receive this message and sample it using their *Camera*. The captured image sequence is processed by the *Image Processor* to restore the encoded bit sequence from the image. Afterwards, it is decoded by the *Channel Decoder* to get the reference time. If the message is reconstructed without any errors, the AC's local clock is set according to the content of the message by the *Local Clock* component.

Receiver Component The ACs in visual range sample the beacon's time-dependent rectangular brightness signal by capturing images with a specific frame rate. In contrast to the traditional sampling problem of analog signals, additional difficulties arise. On the one hand, the rectangular signal makes way for stable signal levels, allowing for robust sampling and so alleviating the decoding process. On the other hand, despite the advantage of stable signal levels, rectangular signals have the obvious disadvantage of containing extremely high frequencies (infinite in theory) each time the signal state changes. Thus, the sampling theorem – which would provide a sufficient condition for perfect signal reconstruction [68] – is not useful in this case, since the image sensor samples its environment with a lower frequency, i.e. the frame rate.

In addition, cameras are not ideal sampling systems, since they do not return the value of the sampled signal instantaneously, but collect light during an exposure time. Nevertheless, the concept of exposure time can be used to reduce the camera back to an ideal sampling system. This is done by the *Image Processor* component. Given the constants c_1, c_2, c_3, c_4 and the intensity of the rectangular signal generated by the beacon $s(t)$, it is assumed that the following three conditions hold at all points of interest in time t:

1. the rate of signal photons emitted by the beacon is $\phi_e(t) = c_1\, s(t) + c_4$,

2. the rate of signal photons per time arriving at the image sensor is $\phi_s(t) = c_2\, \phi_e(t)$,

3. the brightness of the pixel values recorded by the image sensor during a time interval $[t, t+T]$ is proportional (with factor c_3) to the amount of photons hitting the sensor during that interval.

Based on these assumptions, it is possible to transfer the rectangular bright-

5.3. ACTIVE FRAME SYNCHRONIZATION

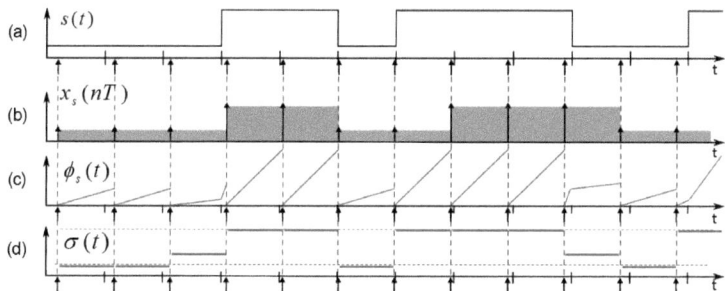

Figure 5.10: The effect of exposure time on signal sampling: (a) original signal generated by the beacon sampled at 103.77% its clock rate (arrows, see text for details); (b) sample values (black), illustrated as zero-order hold values (gray); (c) pseudo signal as per Equation 5.6; (d) sampled pseudo signal as in (b) with high and low reference levels (dashed)

ness signal into the following more robust signal form:

$$\begin{aligned}
\sigma(t) &= f\left(\int_{t}^{t+T} c_3 \phi_s(t) \, dt\right) + n(t) \\
&= f\left(\int_{t}^{t+T} c_2 c_3 \phi_e(t) \, dt\right) + n(t) \\
&= f\left(\int_{t}^{t+T} c_1 c_2 c_3 s(t) + c_2 c_3 c_4 \, dt\right) + n(t) \\
&= f\left(\int_{t}^{t+T} as(t) + b \, dt\right) + n(t)
\end{aligned} \qquad (5.6)$$

The exposure time is modeled by integrating the amount of light on the image sensor over the exposure time T. $s(t)$ is the light energy reaching the image sensor at a given time t. $n(t)$ is an additive noise term for the image sensor. $f(x)$ is an image sensor-specific function considering non-linearities like saturation. While at a first glance, the signal distortion caused by the exposure time may appear detrimental to the decoding process, the new signal form makes way for a more robust sampling. For instance, the robustness for signal sampling and decoding is improved by the low pass filtering effect of the σ function (see Equation 5.6), allowing the complete reconstruction even in the presence of variations of the crystal frequency due to noise, temperature, aging, voltage change, etc. resulting in clock drift. This is illustrated by the example shown in Figure 5.10(a), where a signal is sampled at a sample rate of 1.0377 times the signal clock rate caused by variations in the crystal frequency. As a consequence, two low bit appear in the sampled signal $x_s(nT)$ in Figure 5.10(b) at the sixth and seventh sample point, although there is only one in the original signal $s(t)$. $\phi_s(t)$ is created in Figure 5.10(c). Integrating and sampling $\phi_s(t)$ - based on the assumption of low dynamic ranges of the LED and a noise-free exposure process - leads to $\sigma(t)$ as depicted in Figure 5.10(d) with values anywhere in the range between high and low reference values, as indicated by the horizontal dashed (red) lines.

High and low reference levels have to be generated at runtime, since there is no reference signal level available. In traditional electronic systems, for example, these high and low signal levels are defined at different voltages. Depending on lighting conditions and the current perspective, the same optical bit can appear completely different in terms of its dynamic range and brightness considering only raw values extracted from an image sequence. Therefore, the high and low reference levels have to be reconstructed from series of samples of nearly equal brightness at runtime. The three most recently captured brightness values x_t, x_{t-1}, x_{t-2} and the mean of the N most recent values $m = (x_t + \ldots + x_{t-N+1})/N$ are used to derive reference values from the input stream at runtime (the value of N depends on the specific scenario; we set it to $N = 25$ to consider the values of the last second). If $x_t \geq x_{t-1} \leq x_{t-2}$ and $x_t < m$ then x_{t-1} is considered a low value and added (with weight μ) to the previous estimate for the low reference σ_{low} (weight $1-\mu$) (in our scenario, we set $\mu = 1/3$ to set a higher priority on previous estimates). Analogously, local maxima above m are weighted into the existing high reference estimate

5.3. ACTIVE FRAME SYNCHRONIZATION

σ_{high}.

The *Channel Decoder* reconstructs the original signal from the $\sigma(t)$ representation. For this purpose, we have to differentiate between runs of high values σ_{high} and runs of low values σ_{low}. First, the sum below the curve of $\sigma(t)$ is computed for nT between the start of the rising edge of the run and the end of the falling edge. Calculating the beginning and end of this time range is possible by examining the relative changes of the sample values. For instance, the third sample in Figure 5.10 (d) is higher than the second, hence it marks the beginning of a run of high values. Equally, the sixth sample marks the end of the high run and thus the last of the $n_s = 3$ samples have to be included. The value $n_s \cdot x_{\text{low}}$ is subtracted from the sum. The result, divided by the dynamic range $\sigma_{\text{high}} - \sigma_{\text{low}}$ and rounded to the nearest integer, is the number of clock cycles of the sequence n_{clk}.

Sender Component From the requirements and limitations of the receiver component described so far, a number of constraints for the sender component can be derived:

1. **Signal Changes**: The image processing algorithm needs several intensity changes per second in order to locate and track the signal source.

2. **Peak Values**: In order to reconstruct high and low reference values, occasional peak values are needed.

3. **Run Length**: Noise and variations of the crystal frequency due to noise, temperature etc. make the detection of long runs of the same signal state difficult.

Constraints 1 and 3 are essentially equivalent in the sense that long sequences without a state transition must be avoided. The exact value of the length limit depends on the environment of the system, such as dynamic range of the signal and non-signal activity in the captured scene, but lower maximum lengths increase robustness in any case. On the other hand, condition 2 demands a minimum length of sequences of the same value. In particular, when sampling a $\sigma(t)$-filtered signal, the intensity level has to be constant for

CHAPTER 5. DYNAMIC RECONFIGURATION METHODS

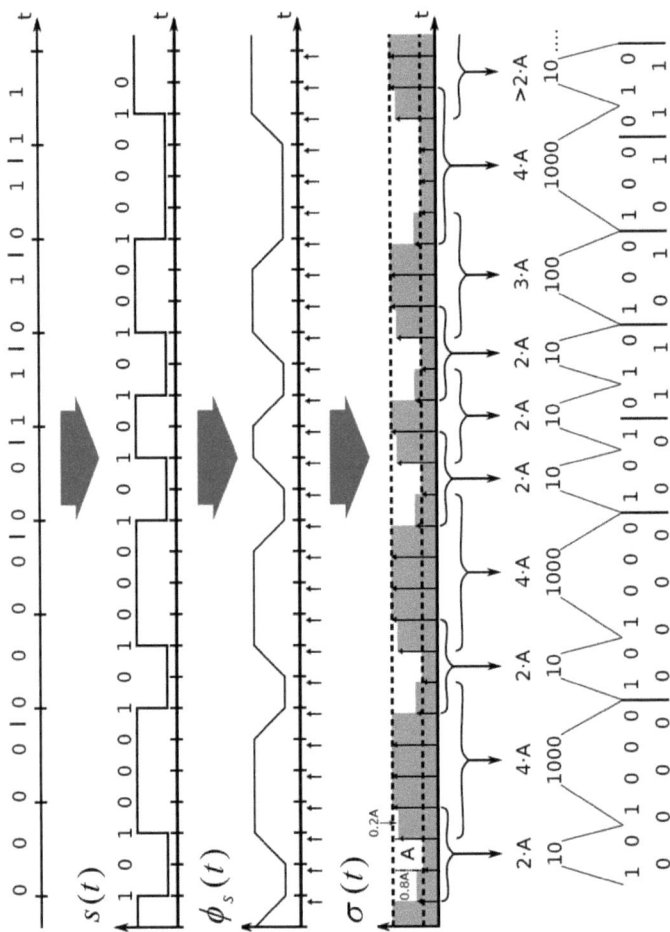

Figure 5.11: The complete sampling process illustrated, apart from the adjustments necessary to account for the physical properties of image sensor saturation.

5.3. ACTIVE FRAME SYNCHRONIZATION

at least two clock cycles in order to guarantee at a minimum one sample at the extreme value.

Codes that guarantee a minimum and maximum run length have been subject to research for roughly 40 years and are called (d, k) run-length limited (RLL) codes [69]. A (d, k) RLL code ensures that there are at least d and at most k *zero* bits between two *one* bits. One common example of a (2,10) RLL code is the eight-to-fourteen modulation from the Compact Disc standard that encodes eight data bits into fourteen channel bits plus three padding bits [70]. In order to increase the efficiency of such codes, RLL encoded bit streams are typically modulated with the *non-return to zero inverted* (NRZI) modulation that generates a state transition for each *one* bit and keeps the current signal level on a *zero* bit. When interpreted this way, the RLL code ensures that the time between two edges in the signal is at least $d + 1$ and at most $k + 1$ clock cycles in every case [71].

To select a code for the system developed in this thesis, three factors need to be considered. First, $d \geq 1$ in order to meet condition 2 stated at the beginning of this section. Secondly, k should be reasonably small. And thirdly, the number of data bits per channel bit r should be reasonably high. Since the last two properties are contradictory, a compromise is necessary. One code that meets the requirements sufficiently is the $(1, 7)$ RLL code [71]. It has a code rate of $r = \frac{2}{3}$, encoding every two data bits into three channel bits according to Table 5.1. Since the code is not prefix-free, it is important to check the four bit substitution table starting from the bottom. If no match is found, the two bit replacement from the top four rows of the table are chosen. This is necessary to ensure the $d = 1$ property, as, for instance, encoding *1001* solely with the standard encoding table would yield *001100* [71].

The entire coding and decoding process is summarized in Figure 5.11.

A synchronization marker (01 0000 0000 10) is added to indicate the start of a code word. The first and last two bits serve two purposes: On the one hand, the two *one* bits define the exact length of the sequence of 8 *zeros*; and on the other hand, the zero at the very beginning and end ensures that codewords ending or beginning with a *one* do not violate the $d = 1$ condition at the transition to the synchronization marker.

Data Bits	Channel Bits
00	101
01	100
10	001
11	010
00 00	101 000
00 01	100 000
10 00	001 000
10 01	010 000

Table 5.1: One possible encoding table for a $(1,7)$ run-length limited code.

Algorithm 8 Beacon Algorithm (operation mode 1)

Require: Beacon has synchronized time
1: init:
2:   $run \leftarrow true$
3:
4: **while** $run == true$ **do**
5:     $sec \leftarrow$ readSecondsSinceMidnight()   // *local time*
6:     **if** $ts \mod 2 == 0$ **then**
7:       $ts \leftarrow sec/2$
8:       $o_{t_0} \leftarrow$ encode(ts)   // *adding $CRC8$ and synchronization marker*
9:       opticalSend(o_{t_0})
10:     **end if**
11: **end while**

Algorithm This subsection presents the algorithms for the aforementioned receiver and sender components. As depicted in Figure 5.12, the beacon (sender component) transmits one time stamp t_0 in such a way that the beginning of the first bit marks time t_0 for which the time stamp is valid. The AC (receiver component) measures the complete duration Δt of the transmission (with some error e) and then adjusts the received time stamp by this measured Δt.

Subsequently, a protocol is presented that transmits an optical message containing a time stamp every two seconds. At a granularity of two seconds per time stamp, a day can be encoded using 16 bit to encode the 43,200 possible values (43,200 corresponds to the number of seconds passed since midnight

5.3. ACTIVE FRAME SYNCHRONIZATION

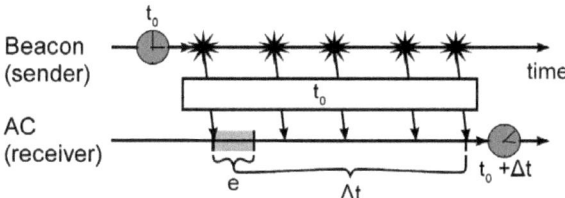

Figure 5.12: Concept of the beacon-assisted clock synchronization algorithm based on the receiver and sender component

divided by two). Nevertheless, this method can be adapted for higher or lower granularities. Taking a granularity of two seconds and assuming a frame rate of 25 fps, the resulting 16 bits take up 24 of the 50 clock cycles available in two seconds. Adding the synchronization marker (01 0000 0000 10), requires another 12 cycles. This leaves 14 clock cycles – equivalent to 8 bits and 2 cycles – which are used to store an 8 bit cyclic redundancy check (CRC) digest. The remaining 2 clock cycles are integrated into the synchronization marker by appending two zeros at the end.

The beacon algorithm is presented in Algorithm 8. The time stamp is – as discussed above – equal to the seconds passed since midnight of the current day, divided by two (cf. Algorithm 8 line 7). For the ease of use, the time stamp is based on the local time instead of UTC. Nevertheless, this can be modified for future versions. The high byte of the 16 bit word is sent first, the low byte second.

Before processing the optical message, the beacon signal has to be found. This is performed by a dynamic mask creation process (see Figure 5.13), which is responsible for finding the relevant pixels in the imagery and constructing the optical message. Since the details of the underlying algorithms are beyond the scope of this thesis, the reader is referred to [8]. The output of this process is the *opticalReceive* event, which is used in Algorithm 9.

The proper error detection method uses an eight bit cyclic redundancy check (CRC8) of the 16 data bits with the generator polynom 0x97 [72]. The resulting checksum is appended directly to the packet, without further modi-

---
Algorithm 9 Beacon-AC Algorithm (operation mode 1)

Require: AC has been repositioned and possesses free view on the beacon signal, beacon signal is found
1: **on** opticalReceive(o_{t_0}) :
2:   **if** isCRCValid(o_{t_0}) **then**
3:     $t_{\text{loc}} \leftarrow$ convertTS(o_{t_0})   // *transform time stamp to time*
4:   **end if**
5: **end on**
---

fications (cf. Algorithm 8 line 8). Validating the code on the AC is conducted by running the same calculation on the first two received bytes and comparing the result to the third byte; on mismatch the received message is discarded (cf. Algorithm 9 line 2).

However, as with all digital signature schemes, there is a small, but finite, probability that a data corruption that inverts a sufficient number of bits in just the right pattern will occur and lead to an undetectable error [72]. While this encoding uses only 43,200 discrete numbers and does not exploit the whole range of 16 bit integers, it allows for an additional layer of error detection at the receiver side: should a faulty transmission yield a correct checksum but a time stamp over 43,199, it is certainly invalid.

Since the beacon-assisted clock synchronization requires the deployment of infrastructure (i.e. beacons) within the workspace, the following section presents an approach that is independent of such an infrastructure in order to increase the overall method's robustness.

5.3.4 Cooperative Frame Synchronization Algorithm

This section explains the implementation of operation mode 2 for cooperative frame synchronization. It does not require specific infrastructure like optical beacons for synchronization, since it is based on the observation of visual events. Each AC passively keeps track of activity in its environment and calculates saliency values based on visual changes. This saliency value is based on the computation of the optical flow vector field (see Section 2.4.2) between the most recently captured frame and the previous one. If this saliency value

5.3. ACTIVE FRAME SYNCHRONIZATION

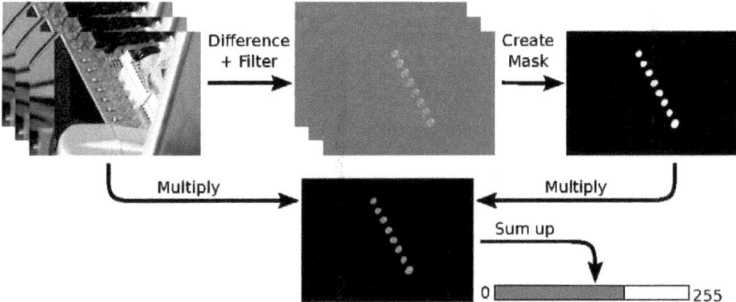

Figure 5.13: The mask creation and multiplication technique. A mask is constructed from a series of frame differences, multiplied with the corresponding frame and then aggregated

increases noticeably, the AC interprets that frame as a salient event and communicates this conclusion to its visual neighbors, along with a copy of its recent saliency history. The neighboring ACs will then correlate the received curve with their own and determine the frame offset at which the correlation is at its maximum, see Figure 5.14. This frame is then registered as the potential clock difference between the receiving camera and the originator of the curve.

First, we will present the methodology for calculating the saliency values based on moving targets within the AC's field of view. Secondly, the algorithm for determining the frame offset in a cooperative manner is explained.

Visual Saliency of Moving Targets The core aspect of coping with visual saliency of moving targets is to detect and quantize them. A moving target is characterized by its high variation in space in one frame to its successive frame. These motions are mostly unique between two frames and may pronounce themselves from different viewpoints. A low-level computer vision algorithm that utilizes adjacent frames for motion estimation is the *optical flow* algorithm [35]. Motion estimation is a major aspect of optical flow research. The optical flow field is superficially similar to a dense motion field derived by techniques like motion estimation. In addition, optical flow is a common technique used

106 CHAPTER 5. DYNAMIC RECONFIGURATION METHODS

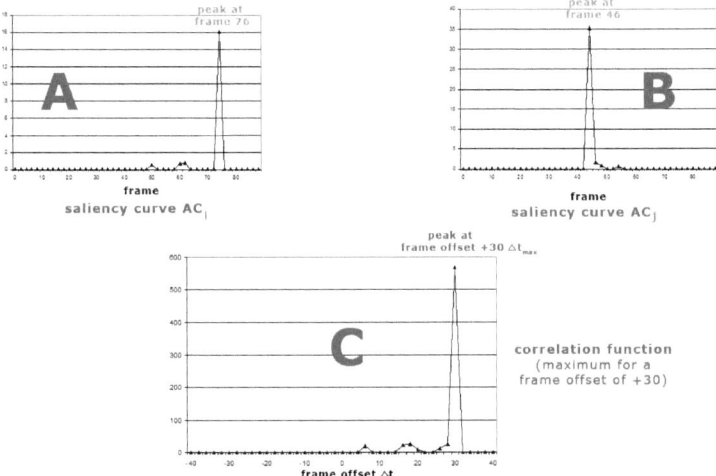

Figure 5.14: Example for calculating the frame offset by correlation

5.3. ACTIVE FRAME SYNCHRONIZATION

Figure 5.15: Spatial offsets of a person entering the camera's field of view

in active vision research and available on many platforms.

The basic approach to calculate the optical flow is to compute the spatial offset between so-called *feature points* of the current image and a previously captured one. Feature points are characteristic high-frequency information such as lines or corners within the image. The spatial offset shows where each feature point in the original image moved to, see Figure 5.15. Based on the *Lucas-Kanade* algorithm [73], the absolute value of the spatial offset so_p^t of a feature point p at time t (e.g. p_y^{t-1} denotes a feature point p at time $t-1$ for the y-dimension) can be derived as follows:

$$\left\|so_p^t\right\| = \sqrt{(p_y^{t-1} - p_y^t)^2 + (p_x^{t-1} - p_x^t)^2} \quad (5.7)$$

For a frame c_t and its preceding frame c_{t-1}, the absolute values of all feature points are clustered into groups and the histogram h_t is created. h_t is the basis for forming the frame-based optical flow distribution $P(h_t)$. $P(h_t)$ and $P(h_{t-1})$ are the histograms' relative frequencies of absolute values at the specific times t and $t-1$. $P(h_t)$ and $P(h_{t-1})$ are used to compute a time-dependent saliency measure $S_j(t)$ for each AC_j at time t:

$$S_j(t) = S_j(c_t, c_{t-1}, c_{t-2}) = KL(P(h_t), P(h_{t-1})) = P(h_t) log \frac{P(h_t)}{P(h_{t-1})} \quad (5.8)$$

KL (the Kullback-Leibler divergence, also called *relative entropy*) measures the difference between the histograms $P(h_t)$ and $P(h_{t-1})$. E.g. if there is no difference between both histograms, the fraction $\frac{P(h_t)}{P(h_{t-1})}$ will be one and $\log(1)$ is zero.

We define the KL of two histograms as follows: for each entry of the numerator and denominator, the normalized value p and q is computed, respectively (whereas the normalization is achieved by dividing the value by the sum of all entries in the histogram). p represents values of the numerator and q of the denominator. Afterwards, the distance $dist = p \cdot (\log(p) - \log(q))$ is computed. The sum of all of these distances is the KL of both histograms.

Storing the values of the saliency measure allows for creating a saliency curve for each AC. The saliency curves of two neighboring ACs, AC_i and AC_j, sharing the same field of view can now be correlated as follows:

$$\Psi_{ij}(\Delta t) = \frac{1}{2T} \int_{-T}^{T} S_i(t) \cdot S_j(t + \Delta t) \qquad (5.9)$$

The frame offset Δt (see Figure 5.14) is an integer within the interval $[-d+1, d-1]$. d is the number of correlated frames. It is defined by the time interval (e.g. 2 seconds) that is to be correlated multiplied by the frame rate. Δt is defined as follows, that it is positive if AC_i lags behind and vice versa if AC_i runs ahead. The Δt at which our correlation function reaches its maximum, is the estimation of the frame offset $\Delta t_{max} = argmax \Psi_{ij}(\Delta t)$.

Algorithm The algorithm consists of the local saliency curve computation and the cooperative correlation of saliency curves of neighboring ACs.

The saliency measure is computed locally as described in Algorithm 10. In each frame, see Algorithm 10 line 6, AC_j computes the saliency measure with the most recent two frames and stores the last $d = 50$ calculations (FIFO buffer of size d, $d = 2\,s \cdot 25\,fps$). If an optical event occurs and a specific threshold δ_s value is reached (δ_s depends on the scenario and is set according to which kind of spatio-temporal pattern is salient in the setting), see Algorithm 10 line 9, AC_j sends its saliency curve (including time stamps for each saliency value) to AC_i using a synchronization message, see Algorithm 10 line

5.3. ACTIVE FRAME SYNCHRONIZATION

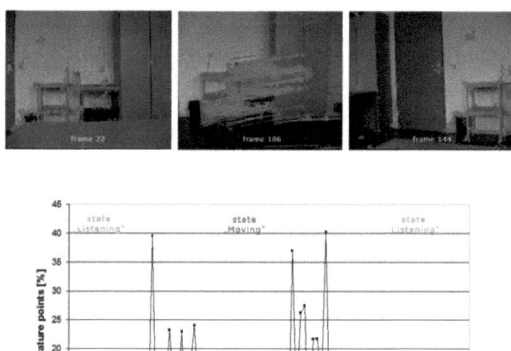

Figure 5.16: Movement detection of an Active Camera

Algorithm 10 Local Algorithm of AC_j (operation mode 2)

Require: AC has at least one neighboring AC with same field of view
1: init:
2: $\quad c_{t-2} \leftarrow empty$
3: $\quad c_{t-1} \leftarrow empty$
4: $\quad V^j \leftarrow empty$ // array for storing saliency values
5:
6: on captureFrame(c_t):
7: $\quad$ if $c_{t-1} \neq c_t$ then
8: $\quad\quad V^j \leftarrow V^j \oplus S_t^j(c_t, c_{t-1}, c_{t-2})$
9: $\quad\quad$ if $S_t^j(c_t, c_{t-1}, c_{t-2}) > \delta_s$
10: $\quad\quad\quad sendSyncMessage(V^j)$
11: $\quad\quad$ end if
12: $\quad\quad c_{t-2} \leftarrow c_{t-1}$
13: $\quad\quad c_{t-1} \leftarrow c_t$
14: end if

10. The step of computing the saliency curve for a frame is dominated by the image size, since the computational complexity of the optical flow algorithm is proportional to the image size, i.e. $O(width * height)$. The next step of calculating the correlation is dominated by the number d of frames, which are to be correlated, since the complexity of our correlation is $O(d^2)$. Experiments show that measured times are feasible as presented in the Section 6.4.6.

Algorithm 11 AC-AC Algorithm (operation mode 2)

Require: AC_j and AC_i share the same view the last 2 seconds (state Listening)
1: on receiveSynMessage(V^i):
2: **if** isNeighbor(i) **then**
3:    sendSyncMessage(V^j)   // V^j is local visual saliency curve
4:    $\Delta t_{max} \leftarrow argmax \Psi_{ij}(\Delta t)$
5: **end if**

Since it is very difficult to detect salient events whilst ACs move, ACs have to share the same field of view for at least 2 seconds. Relative movements distort the optical flow measurement process. Since ACs' movements are a very important aspect of Active Camera Networks, two states haven been introduced for an AC: *Moving* and *Listening*. If an AC is in state *Moving*, it does not search for salient events. This state can be recognized by sensors like accelerometers or visually by calculating the loss of feature points (i.e. difference of the number of retrieved feature points in two consecutive frames) in the camera's field of view. The optical flow measurement gives results about the movement's velocity and change of direction, see Figure 5.16. If the loss is constantly below 5 % (i.e. for more than 2 seconds, e.g. 50 frames), we assume that the AC's movement is over and change to the state *Listening*. This state is marked with A in Figure 5.16. In this state, the AC tries to detect salient events. The state *Moving* is marked with B in Figure 5.16. Here, a loss of feature points is observed, which is above a specific treshold (e.g. > 5%, marked by the red line). In state *Listening* and at arrival of a synchronization message, AC_j sends its local saliency curve to the sender and starts a correlation in order to determine its own frame offset Δt_{max}, see Algorithm 11 line 4, which is used for correcting the AC's local clock.

5.4 Summary

This chapter contains the description of two dynamic reconfiguration methods for Active Camera Networks.

First, a distributed control algorithm called *DRofACN* was presented, which is able to solve the wide-area target acquisition problem by means of dynamic reconfiguration. *DRofACN* allows for scalable and dynamically self-configurable Active Camera Networks without utilizing a priori information. Thereby, drawbacks of passive camera networks can be overcome. The basic idea of *DRofACN* is that a set of ACs collaborates for acquiring close-up views of targets in a surveillance area. The AC control is based on the output of perceiver nodes, i.e. generated target requests, estimating the position of salient targets within the ACs' actuation ranges. The main goal is to acquire views for object tracking or for biometric purposes. Furthermore, we have presented a mechanism called *ENRA* for paving the way for phenomena adaptivity of the *DRofACN* algorithm. Based on the number of events occurring in the workspace, Active Cameras increase or decrease so-called *spatial redundancy regions* with neighboring nodes to balance the network's load.

Secondly, a reconfiguration method called *ACFSync* for visual event-based frame synchronization has been presented. With *ACFSync*, Active Cameras can utilize spatio-temporal properties of visual events for frame synchronization. *ACFSync* enables participating nodes to synchronize their clocks according to an optical broadcast signal (sender-to-receiver approach) and reason about frame offsets of neighboring cameras sharing the same field of view (receiver-to-receiver approach). *ACFSync* is only based on image information and provides a solution for Active Camera Networks without suffering from synchronization errors due to message-based communication, since image information propagates with speed of light and arrives at the image sensor of cameras sharing the same field of view nearly at the same point of time. Traditional message-based communication is only used to exchange status messages but not for extracting timing information for time synchronization.

The following chapter contains an evaluation of both reconfiguration methods.

Chapter 6

Evaluation

This chapter is devoted to examine the performance of the proposed architecture and reconfiguration methods. Initially, relevant metrics that allow to measure and analyze the system's performance are introduced. Afterwards, the conducted experiments are presented. The experiments that comprise the evaluation are derived from application scenarios for Active Camera Networks, e.g. surveillance of public places. The evaluation of the algorithm *DRofACN* has been carried out in a simulated environment, whereas the frame synchronization method *ACFSync* has been evaluated on artificial and real video sequences from multiple cameras. Simulation experiments as well as artificial video sequences allow for the investigation of large networks with tens to hundreds of cooperating Active Cameras.

6.1 Performance Metrics

Distinct metrics can be applied in order to measure the performance of the system presented in this thesis. The overall architecture can be evaluated in terms of scalability, e.g. by increasing the number of Active Cameras (ACs) or targets of interest (ToIs) within the surveillance area. For the frame synchronization method *ACFSync*, the synchronization accuracy is of major interest, i.e. how accurate the camera's clock can be synchronized on the basis of visual data.

Each of the following subsections contains a short introduction to the investigated reconfiguration method and a number of performance metrics. Each subsection closes with formulating research questions, which are answered in this chapter.

Performance Metrics for DRofACN Several metrics have been applied to measure the performance of the *DRofACN* method, which is a heuristic for the wide-area target acquisition problem introduced in Section 5.2:

- **Target acquisition ratio (TAR):** The quality of the dynamic reconfiguration is described by the target acquisition ratio (TAR), whereas the minimum imaging quality $q_{\min}$ is set to 0.75 (0.75 means that a reconfiguration has to be chosen possessing a computer vision success rate of at least 75%). A TAR ratio of one means that every target entering the surveillance area is observed before leaving. According to the definition in Equation 5.1, the target acquisition ratio TAR is defined as the ratio of successfully observed targets to the total number of targets entering the surveillance area in the time period $[t', t'']$, i.e.: $TAR = \frac{1}{n}\sum_{t=t'}^{t''} \sum_{\text{target}_i \in \text{ targets}} \text{target}_i.observed(t)$

- **Mean target detection time:** Another metric is the mean target detection time. This metric measures how much time is needed to acquire a picture of the target after becoming a target of interest.

- **Captures per target:** The number of captures per target is measured. This metric is important to investigate how far the observation condition is fulfilled in case of disturbances such as packet loss.

- **Load:** Finally, the system's load is measured by investigating the total non-idle time of all ACs. If an AC is not in the mode *IDLE*, it is active and maximizes the system's performance. Therefore, this metric is important to measure the system's overall utilization.

The following research questions are to be answered in this chapter:

6.1. PERFORMANCE METRICS

- How scalable is the system in terms of the number of nodes and the target generation rate? The target generation rate describes the number of targets entering the surveillance area.

- How robust is the system towards localization uncertainty/errors of the perceiver nodes? Perceiver nodes are prone to errors, since they estimate the targets' location on the basis of low-cost sensor technology such as motion detectors.

- How well can the system cope with distinct types of trajectories? Human targets may not always move on a straight line due to pathways or obstacles.

- How does the target speed influence the system's performance? Usually, humans move with a velocity of $1.5 \frac{m}{s}$. Nevertheless, running humans or bicyclists may move with a target speed of up to $15 \frac{m}{s}$.

- How does the loss of packets influence the system's performance? Packet loss on the camera-to-camera communication channel may disturb the cooperation among cameras. On the other hand, packet loss on the perceiver-to-camera communication channel leads to a loss of target requests.

Performance Metrics for ACFSync (operation mode 1)

- **Synchronization accuracy:** The synchronization accuracy measures the difference between the encoded time in the time stamp and the point of time at which the AC is able to set its clock. Inaccuracies may arise due to delays on the link between the image sensor and the AC's processing unit.

- **Error rate:** This metric measures the number of successfully decoded packets in relation to the total number of synchronization messages sent by the beacon.

- **Time complexity:** This metric measures the time period needed to detect and decode an optical message.

- **CPU and memory utilization:** This metric is important to verify the algorithm's real-world applicability.

The following research questions are investigated in this chapter:

- How much time does it take to retrieve image data from an IP camera through an Ethernet link, since this influences the synchronization accuracy?

- How does the signal strength, i.e. the amount and brightness of pixels in the video stream stemming from the beacon, influences the number of successfully decoded optical messages? Is the error rate influenced by distinct lighting conditions?

- How long does it take to detect and decode an optical message on a Smart Camera? Is the time complexity influenced by different lighting conditions?

Performance Metrics for ACFSync (operation mode 2)

- **Calculated frame offset:** This metric is used to measure the offsets of the video sequences of neighboring ACs. In relation to the ground truth, i.e. the real frame offset of the video sequences, the accuracy of the method can be examined by correlation.

- **Certainty scale (of frame offset):** This metric measures the ratio between the amplitudes of the two highest correlation peaks A_{peak1} and A_{peak2}, i.e. $cert_{\text{offset}} = (A_{\text{peak1}}/A_{\text{peak2}}) - 1$. A peak is a maximum if there are only smaller values in its neighborhood. If both peaks have the same height, $cert_{\text{offset}}$ is 0. If the first peak is two times higher, $cert_{\text{offset}}$ is 1 etc. This metric is used to illustrate the distance of the calculated frame offset from its nearest neighbor in the result space. Thereby, it indicates the certainty of the calculated frame offset and thus characterizes the method's robustness.

- **CPU and memory utilization:** This metric is important to verify the algorithm's real-world applicability.

6.1. PERFORMANCE METRICS

The following research questions are investigated in this chapter:

- How does the synchronization accuracy depend on the duration of the visual event?

- How does the perspective influence the accuracy of the cooperative frame synchronization mechanism? How do signal distortions, e.g. stemming from varying illumination or JPEG/MPEG coding algorithms etc., influence the accuracy of the cooperative frame synchronization mechanism?

Several parameters influence the performance of the aforementioned reconfiguration methods. All above introduced metrics are measured under the influence of at least one of the following parameters:

- **System size in number of cameras:** The system size influences the communication effort within the network. The more camera are available the more coordination is needed within the network.

- **Number (and dynamics) of targets in the surveillance area:** The number (and dynamics) of targets is responsible for the load induced into the network. The more targets are within the surveillance area the more target requests are generated. The velocity and trajectory of the target determine how much time is available for observing the target while traversing the surveillance area.

- **Message loss of the communication channel:** Message loss of the communication channel is an important disturbance in terms of the reconfiguration methods, since the communication between neighboring nodes is restricted. Thereby, they are less able to cooperate on target observation.

- **View angle and distance of the camera capturing imagery / Signal strength (signal area in image and brightness):** Both parameters have strong influence on the quality of the underlying image processing algorithms.

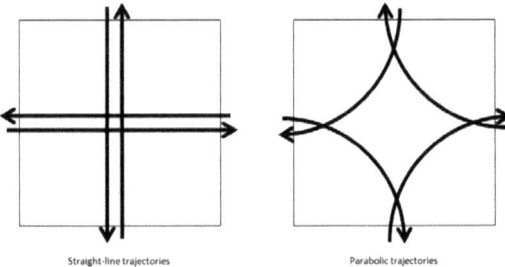

Figure 6.1: Experiments with targets entering the surveillance area on straight-lines (left picture) or parabolic trajectories (right picture)

6.2 DRofACN

6.2.1 Experimental Setup

Prior to the presentation and discussion of experimental results, the experimental setup is introduced.

The area to be observed is 250×250 square meters. This is a common size for public places, e.g. such as the front yard of the Hannover Main Station as depicted in Figure 1.1. ACs are positioned on a grid of this area, whereas the actuation ranges of neighboring ACs overlap by a quarter of the actuation radius (as depicted in Figure 6.2). The system size is varied between $1 \times 1, ..., 10 \times 10$ ACs. The actuation radius of each camera is calculated as follows: $a_r = \frac{1}{2} \cdot 250m/camsInRow$. E.g. in a 10×10 ACs scenario, 10 cameras are in a row. Thus, the actuation range is $a_r = 12.5\,m$ in this case. The actuation range decreases with the number of ACs in order to support dynamic reconfiguration. Nevertheless, the number of spatial conflicts (see Section 5.2.6) and communication per area is increased.

ToIs entering the surveillance area are generated with a specific target generation rate. This rate varies between $0.05\,ToIs/s, ..., 27.5\,ToIs/s$. Since the ToIs move with a speed of $1.5\,\frac{m}{s}$, they need approx. $166\,s$ traversing the surveillance area in case of a straight-line trajectory. Thus, a rate of

6.2. DROFACN

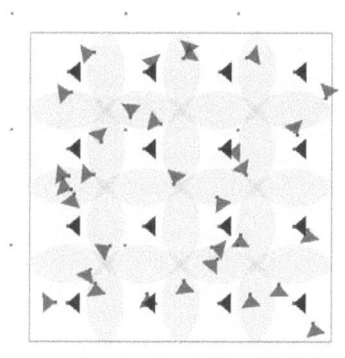

Figure 6.2: ACs are positioned in a grid on the surveillance area. The actuation ranges of neighboring ACs overlap by a quarter of the actuation radius and decreases with the number of ACs.

$0.05\,ToIs/s$ corresponds to a total number of $8.3\,ToIs$ in steady state. In case of a rate of $27.5\,ToIs/s$, up to approx. $4,500$ ToIs are within the surveillance area in steady state. Assuming a surveillance area of 250×250 square meters with a potential capacity of $65,000$ persons (assuming that a person requires space of $1\,m^2$ [74]), a rate of $27.5\,ToIs/s$ means that one out of twenty targets is a ToI. Targets become ToIs on their way crossing the surveillance area. Target requests are sent by the perceiver nodes with a frequency $f_p = 1\,Hz$. A frequency of $1\,Hz$ is a realistic value for perceiver nodes based on off-the-shelf sensors as introduced in Section 2.2. By increasing this frequency, the load within the camera network is increased, too, since the number of generated target requests raises. The point of time, when becoming salient, is distributed over their entire way in a uniformly way. In order to investigate the robustness of *DRofACN*, two different trajectory variations of ToIs have been considered. Usually, ToIs move on a straight-line. Nevertheless, in one experiment they move on a parabolic trajectory. In both cases, ToIs enter the surveillance area from all sides, whereas the entrance point is randomly chosen. An example for both types of trajectories can be found in Figure 6.1. Table 6.1 shows the

	Experimental setup	
	Scalability	Robustness
	System size	System size
Number of ACs	1 ... 100	Number of ACs: 4 ... 36
Target generation rate	$0.5\,ToIs/s$ ... $27.5\,ToIs/s$	Target generation rate: $6\,ToIs/s$
	Packet loss	Packet loss
AC-AC	0 %	AC-AC: 0...100 %
Perceiver-AC	0 %	Perceiver-AC: 0...100 %
	Targets	Targets
Trajectory	straight	Trajectory: straight, parabolic
Localization uncertainty/error h	$\pm 0\,cm$	Localization uncertainty/error h: $\pm 0\,cm ... \pm 1\,m$

Table 6.1: Experimental setup

experimental setup in a compacted form.

6.2.2 Scalability

Scalability with respect to the target generation rate is an important factor for Active Camera Networks. Scalability concerns both the quality of the solution found for the wide-area target detection problem (i.e. resulting in a high TAR and low target detection times) as well as the number of resources needed. Results can be found in Figure 6.3 and 6.4. These figures depict the relation of resources needed, i.e. ACs, to the algorithm's quality of service in terms of the number of observed targets (TAR) and the mean target detection time.

The following example illustrates how these results have to be interpreted: We assume that a surveillance scenario, e.g. a public place of 250×250 square meters, is given where approx. 10 targets enter the surveillance area per second, i.e. a target generation rate of $10\,ToIs/s$. In addition, an image of 90 % of the targets has to be acquired within a mean target detection time of less than 10 seconds. As depicted in Figure 6.3, a system size of 50 cameras is

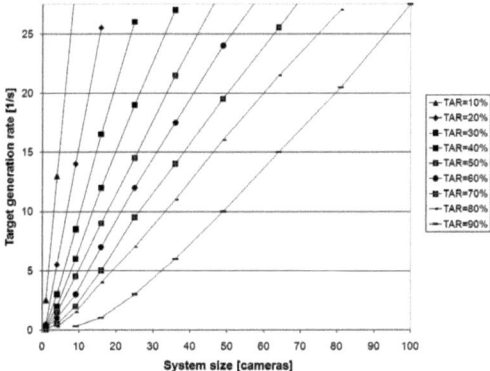

Figure 6.3: Relation of target generation rate and resources needed for achieving a specific TAR ratio: For target acquisition ratios up to 80%, the number of ACs needed increases proportionally to the number of ToIs. For a TAR > 90% significantly more resources are needed.

needed to achieve a TAR ratio of 90% for the given target generation rate of $10\,ToIs/s$. A system of 50 cameras is able to achieve a mean target detection time below 10 seconds for target generation rates below (approx.) $5\,ToIs/s$, see Figure 6.4. Nevertheless, in order to achieve this for a target generation rate of $10\,ToIs/s$, the number of resources within the camera network has to be increased to approx. 80 cameras.

The load induced into the system is depicted in Figure 6.5. A TAR ratio of up to 80% can be achieved by increasing the number of ACs in the network according to the target generation rate. In order to achieve a TAR ratio higher than 80%, the number of resources needed for a specific target generation rate does not increase linearly any more. Here, significantly more resources are needed. The reason is that it must be guaranteed that targets which are about to leave have to be captured, too. This can only be achieved by a high AC density, since only small actuation ranges can guarantee that these ToIs are observed. In order to achieve mean target detection times below 10 seconds, significantly more resources are needed. The target detection

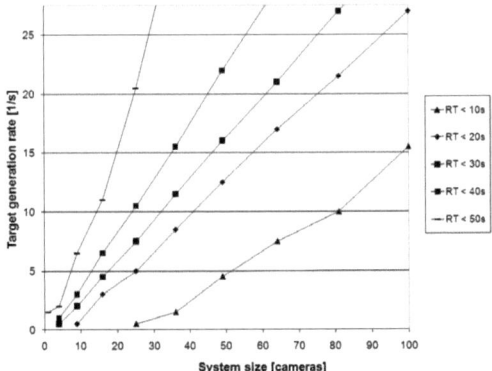

Figure 6.4: Relation of target generation rate and resources needed for achieving a specific mean target detection time: For a mean target detection time below $20\,s$ the number of ACs increases proportionally to the number of ToIs. For mean detection times below $10\,s$ significantly more ACs are needed.

time consists of the time needed to capture and to process the image of the ToI successfully after becoming salient. This time period includes the time for receiving the target request (up to $1\,s$, since perceiver nodes send target requests with a frequency of $1\,Hz$), the time for repositioning the AC (depends on the AC's actuation range), and the time for processing the image ($1,000\,ms = 1\,s$). A mean target detection time of 10 seconds means that the majority of ToIs is detected in the neighboring actuation range at the latest after becoming salient. For this purpose, neighboring nodes must not be in an overload situation. To guarantee this, a high number of ACs is needed.

Additionally, the results show that the imaging quality decreases if the number of ACs is increased, see Figure 6.6. This observation is caused by the decrease of the AC's actuation range, too, which leads to a decrease of the area available for finding the optimal position for image capturing. Nevertheless, the imaging quality increases with increasing load (i.e. target generation rate), as an AC has more choices for observation.

6.2. DROFACN

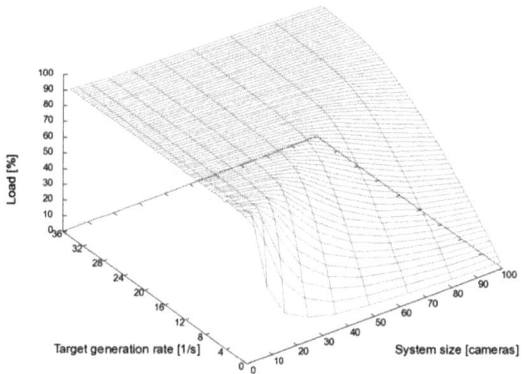

Figure 6.5: System size and influence on load

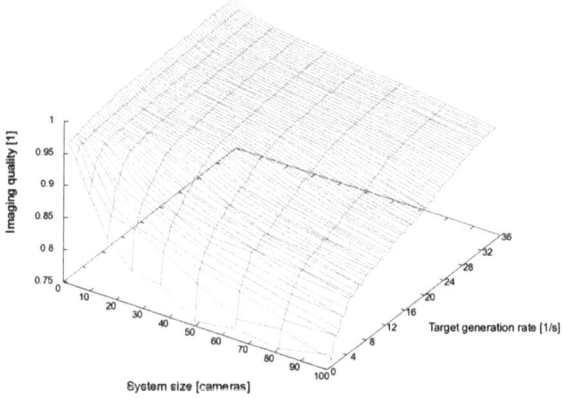

Figure 6.6: System size and influence on imaging quality

124   CHAPTER 6. EVALUATION

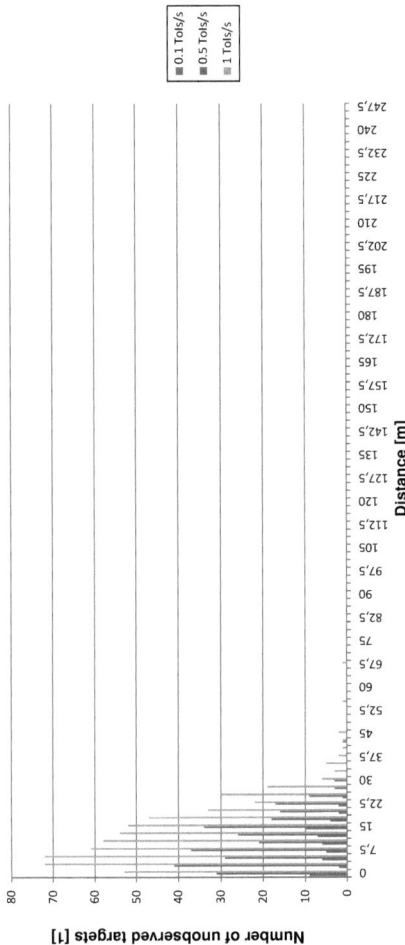

Figure 6.7: Number of unobserved targets for low target generation rates for an ACN consisting of 16 cameras in relation to the target's distance at time of becoming salient before leaving

6.2. DROFACN

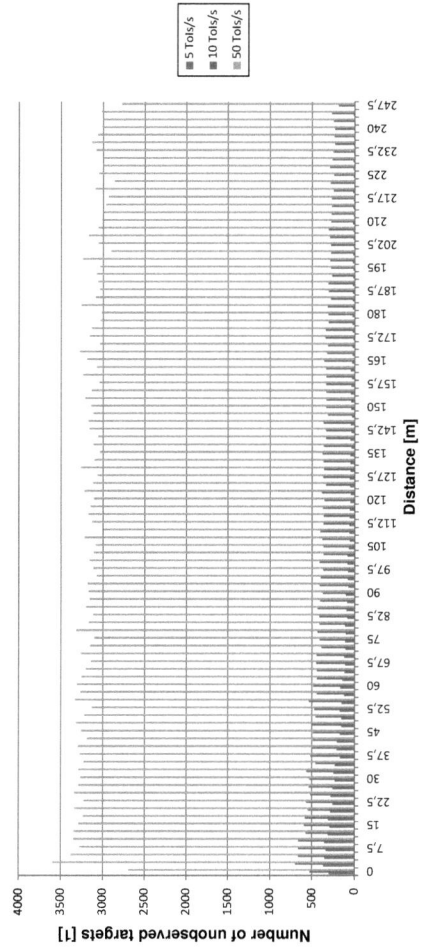

Figure 6.8: Number of unobserved targets for high target generation rates (overload) for an ACN consisting of 16 cameras in relation to the target's distance at time of becoming salient before leaving

Why can a TAR ratio of 100% hardly be achieved? A TAR ratio of 100% is hard to achieve by the *DRofACN* algorithm due to the targets' saliency model. Targets become salient when traversing the ACN's surveillance area. The point of time when becoming salient is distributed over their entire way in a uniformly manner. This means that a target can become salient exactly before leaving the surveillance area. In this case, the AC can only capture an image of the target if it is nearby by accident. Nevertheless, this is very unlikely since the AC can be at any location of its actuation range. In Figure 6.7, for example, the number of unobservable targets (and their distances before leaving when becoming salient) is presented. In case of low target generation rates, i.e. not an overloaded scenario for the ACN consisting of 16 cameras, only targets which become salient approx. 30 meters before leaving cannot be observed. 30 meters is the actuation radius of an AC in an ACN consisting of 16 cameras. These targets cannot be observed if the AC, for example, is situated at the opposite side of its actuation range. This effect may be reduced by increasing the AC's moving speed, which has not been investigated in this thesis.

However, in overloaded scenarios (as depicted in Figure 6.8), the number of unobserved targets (and also the distances to their exit points) increases strongly. The reason is that the *DRofACN* algorithm is not able to schedule the targets due to the high load. This leads to a decrease of the overall TAR ratio in overloaded scenarios.

6.2.3 Packet Loss

Results show that packet loss concerning the AC-to-AC communication has minor impact on the system's performance, i.e. the TAR ratio. Figure 6.9 illustrates the observation. The TAR ratio is only decreased by 10% for scenarios with up to 9 cameras. Nevertheless, packet loss has a higher impact on the TAR ratio, if the number of ACs is increased. For a network consisting of 36 ACs, TAR is decreased by approx. 20%. This is caused by the fact that networks which are comprised of a higher number of cameras are affected in a stronger way, since packet loss has a negative impact on the coor-

6.2. DROFACN

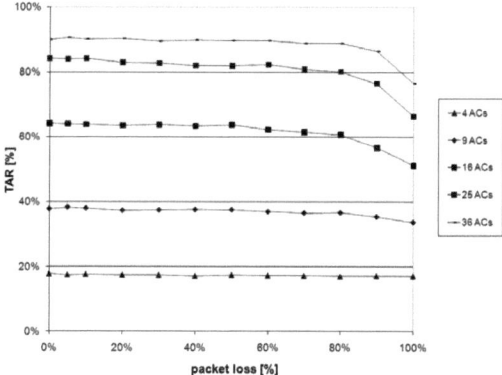

Figure 6.9: AC-to-AC packet loss and the influence on TAR. In case of 100 % of packet loss, the observation condition is not met any more but the TAR ratio only decreases to 80 % of the original value (in scenarios with more than 16 ACs).

dination among the nodes. Thus, scheduling messages (i.e. whether a target has been or is planned to be scheduled) cannot be exchanged between nodes. The results also show that the observation condition is not met any more in case of 100 % of packet loss, since the observations per ToI reach a value of 2.11 $captures/ToI$. This redundant capturing of ToIs causes the TAR ratio to decrease, since resources are wasted due to multiple captures.

Packet loss concerning the perceiver-to-AC communication leads to a decreased system performance, too. In case of 100 % of packet loss, the TAR ratio becomes zero, since no target requests reach their destinations any more. Nevertheless, 90 % of packet loss only decreases the TAR ratio to 70 % of the original value, which proves the robustness of $DRofACN$ towards packet loss. Figure 6.10 illustrates this observation.

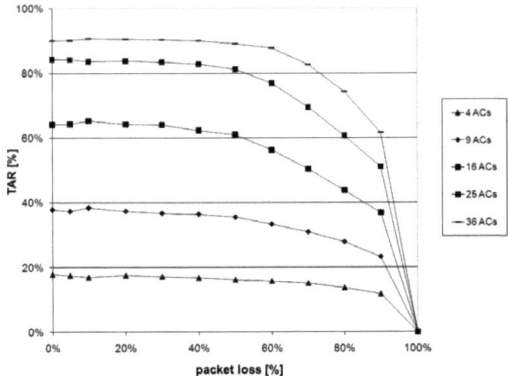

Figure 6.10: Perceiver-to-AC packet loss and influence on TAR. In case of 100 % of packet loss, no target requests reach the ACs any more.

6.2.4 Motion of Targets

Since perceiver nodes estimate the targets' position based on sensor technology, estimation errors become inevitable. In addition, targets may not move on a straight line as assumed in the previous experiments. Since *DRofACN* utilizes the last two positions of a ToI for predicting its future location in a linear way, the target motion may influence the system's performance. As depicted in Figure 6.11, target localization errors of up to $20\,cm$ can be tolerated by the system. The more resources are available, the less is the influence of localization errors on the system's performance. Localization errors of up to $20\,cm$ can arise in perceiver networks consisting of LASER scanners. Localization errors of up to $1\,m$ are common for perceiver networks based on acoustic sensors or motion detectors. In case of localization errors of up to $1\,m$, the TAR ratio is decreased to only 50 % of the original value (for a network consisting of 36 ACs). The impact of localization errors on the target detection time is higher in networks with more than 10 nodes, see Figure 6.12. This is caused by the fact that the ratio of the localization error to actuation range decreases in small networks.

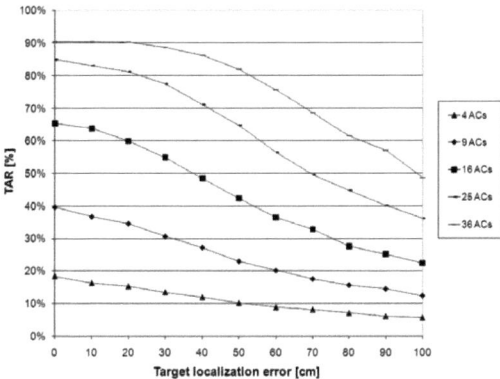

Figure 6.11: Perceiver-target localization error and its influence on the system's performance: An error of up to $20\,cm$ does not influence the TAR ratio significantly.

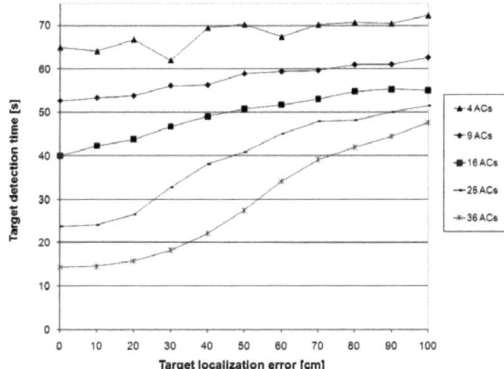

Figure 6.12: Perceiver-target localization error and its influence on the system's performance: An error of up to $20\,cm$ does not influence the mean target detection time significantly.

Parabolic trajectories have a similar effect on the system's performance, since the assumption of linear movement of the targets is violated. Thereby, inaccuracies in terms of extrapolating the target's position occur. In a camera network consisting of 36 cameras, the TAR ratio is decreased to approx. 80 % which corresponds to a localization error of $50\,cm$, see Figure 6.11. This is a reasonable value, since targets move with $1.5\,\frac{m}{s}$ and target requests are generated with a frequency of $f_p = 1\,Hz$. Thereby, depending on the curvature of the trajectory, the localization error is between $0\,m$ and $1.5\,m$. Nevertheless, this effect can be decreased by increasing the frequency $f_p = 1\,Hz$ of generating target requests, since the higher this frequency is the lower is the impact of the trajectory's curvature on predicting the object's future position.

6.2.5 Target Speed

In this section, we investigate how the TAR ratio is influenced by the target speed. In the evaluations in Sections 6.2.2-6.2.4, the target speed was set to $1.5\,\frac{m}{s}$ and the AC speed to $5\,\frac{m}{s}$. In this section, the AC speed is still set to $5\,\frac{m}{s}$ but the target speed is varied between $0.5\,\frac{m}{s}$ and $35\,\frac{m}{s}$ (i.e. approx. $1.8\,\frac{km}{h}$ and $126\,\frac{km}{h}$). The target generation rate is constantly set to $1\,ToIs/s$. As explained in Section 3.2, a perceiver node is deployed at each AC's center of movement with a sensing range defined as twice the camera's actuation range.

As depicted in Figure 6.13, the TAR ratio increases for low target speeds, since the load in the network is decreased. Furthermore, increasing the number of cameras (i.e. resources) in the network leads to an additional improvement of the TAR ratio. Nevertheless, if the target speed exceeds a specific value, namely $8\,\frac{m}{s}$, this effect does not hold any more. In this case, the TAR ratio cannot be improved by increasing the number of resources in the network. Furthermore, for high target speeds ($> 20\,\frac{m}{s}$), an increase of resources within the network decreases the overall performance. This behavior is associated with the sensing range of the ACs. As explained in Section 3.2, a perceiver node is deployed at each AC's center of movement with a sensing range defined as twice the AC's actuation range. Thereby, target requests outside of the region are ignored by the AC, since they are not sensed. By increasing the number

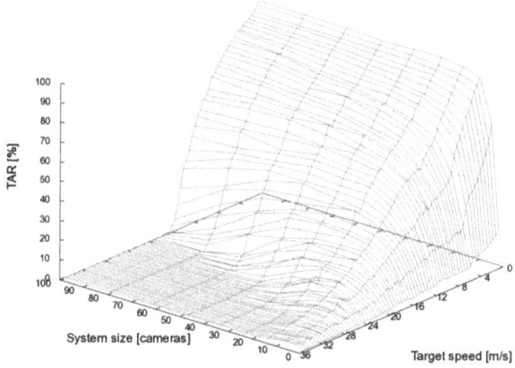

Figure 6.13: Evaluation of TAR in relation to the target speed and system size (sensing range: two times the actuation range).

of cameras within the network, the AC's actuation range is decreased (see Section 6.2.1). The actuation radius of each camera is calculated as follows: $a_r = \frac{1}{2} \cdot 250m/camsInRow$. E.g. in a 10×10 ACs scenario, 10 cameras are in a row. Thus, the actuation range is $a_r = 12.5\,m$ in this case. However, a lower actuation range also means less reaction times for a camera to detect ToIs. Through the low sensing ranges (being twice the actuation range), the cameras notice the ToIs very late and they are missed. Nevertheless, this is no problem for low target speeds (e.g. below $5\,\frac{m}{s}$ as considered in the previous evaluations).

By increasing the sensing range (e.g. setting it to five times the actuation range), the reaction time for the camera is increased. Thereby, the aforementioned problem can be solved for high target speeds as illustrated by Figure 6.14. Nevertheless, by increasing the sensing range, the computational effort of the asynchronous scheduling process to compute the next observation task (see Section 5.2.1) rises. Additionally, it depends on the underlying sensors used as perceiver nodes in how far the sensing range can be increased.

132   CHAPTER 6. EVALUATION

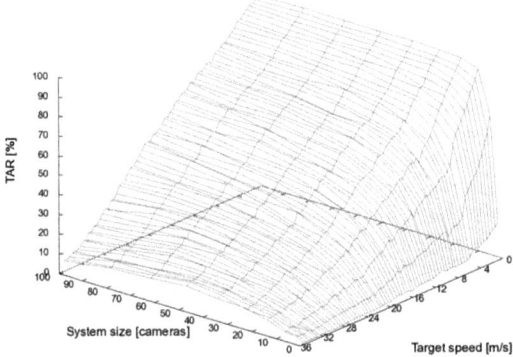

Figure 6.14: Evaluation of TAR in relation to the target speed and system size (sensing range: five times the actuation range)

As explained in Section 6.2.2, the target detection time can be decreased by increasing the number of resources in the network. This effect also holds for high target speeds. Nevertheless, in case the target speed exceeds the velocity of the ACs ($> 5\frac{m}{s}$), a static delay (up to 5 seconds for a camera network consisting of only one camera) is added to the overall detection time. This is illustrated by Figure 6.15.

Additionally, overload leads to an overall decrease of the mean target detection time, since the cameras have to perform less reconfiguration for capturing images of targets. Whether a low mean target detection time is evoked by an overload situation or sufficient resources, is indicated by the reaction time's standard deviation. Table 6.2 shows the corresponding values for the standard deviations of Figure 6.15 of a network consisting of 100 cameras.

6.2. DROFACN

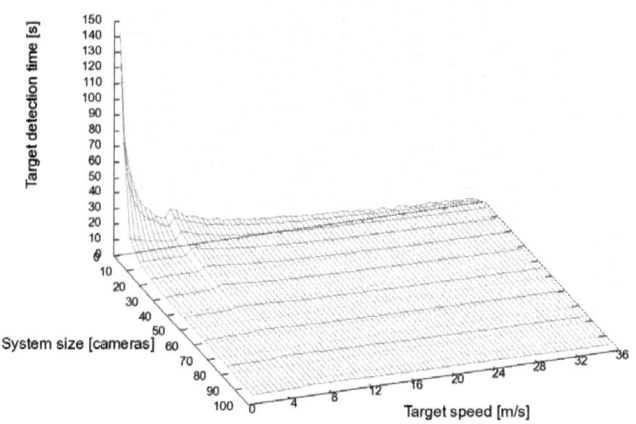

Figure 6.15: Evaluation of the target detection time in relation to the target speed and system size (sensing range: five times the actuation range)

Target speed	TAR [%]	Target detection time [s]	
		Average	Standard deviation
0.5 m/s	97.9781	6.1686	1.1510
1.5 m/s	95.3780	6.1937	1.2782
5 m/s	84.6782	6.6133	2.0729
7 m/s	79.6277	7.4134	2.1878
15 m/s	61.6986	6.3644	1.8380
20 m/s	47.7547	6.0576	1.6548
25 m/s	31.1684	5.7385	1.5612
30 m/s	14.1184	5.3744	1.3705
35 m/s	05.4181	4.9492	1.2182

Table 6.2: Mean target detection time and standard deviation for different target speeds (target generation rate of $1\,ToIs/s$ and 100 ACs; sensing range is five times the actuation range)

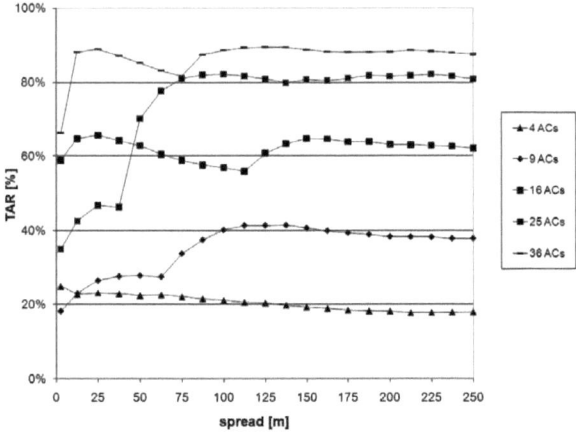

Figure 6.16: System performance (TAR) without network reconfiguration (target generation rate: $6\,ToIs/s$; the spread represents the trajectory's width and targets enter the surveillance area uniformly distributed over this width)

6.2.6 Phenomena Adaptivity (ENRA)

In this section, we present the results of the evaluation of the mechanism for phenomena adaptivity, which has been presented in Section 5.2.6. We conducted experiments for different node placements and spreads. Figures 6.16 and 6.17 show that the TAR ratio can be improved significantly in case of trajectories with small spreads. The node distance n_d is set to 1.95, i.e. ACs are able to select the optimal center of movement in an area having a size of $\pi \cdot [(1+n_d)a_r]^2$. The reconfiguration process is executed every 10 seconds, i.e. $TTU = 10\,s$. The spatial density of the trajectory is defined by its spread. The *spread* represents the trajectory's width and targets enter the surveillance area uniformly distributed over this width. In our experiment, the trajectories are straight-line trajectories entering the scenario from east, west, north and south as depicted in Figure 6.1. The spread of the trajectories varies between $2.5\,m$ up to $250\,m$.

The improvement for common path widths of trajectories induced by hu-

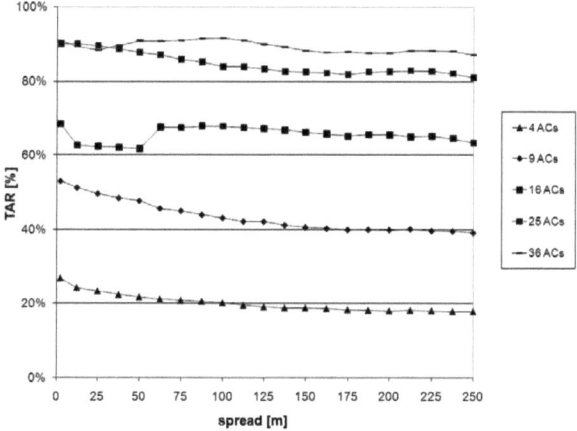

Figure 6.17: System performance (TAR) with network reconfiguration (target generation rate: $6\,ToIs/s$; the spread represents the trajectory's width and targets enter the surveillance area uniformly distributed over this width)

mans, e.g. $2.5\,m$ up to $12\,m$, is much higher than for spreads of $100\,m$ up to $250\,m$. This demonstrates the real-world applicability of our algorithm.

By decreasing the spread of the trajectories, the target density is increased. The reason is that the same number of targets is distributed over a smaller area. Despite this fact, *ENRA* is able to increase the system's performance for spreads below $75\,m$. As the spread of the trajectories decreases, the ACs have to reconfigure their center of movement to achieve the optimal position and generate new spatial redundancy regions to balance the load. These redundancy zones make way for cooperation between nodes and increase the TAR ratio. On the other hand, for widths over $75\,m$, the scenario is overflowed with targets, and the nodes do not need to reconfigure their center of movement. For these values, the improvement achieved by *ENRA* is small. Nevertheless, this improvement is driven by collaboration between neighboring nodes. In case the trajectory's spread is increased, cooperation is decreased.

Additionally, as illustrated in Figure 6.17, camera networks having an odd number of elements achieve higher performance improvements. For camera networks of 3x3 and 5x5 cameras, the node placement is centered in the surveillance area. Thus, the ACs in the center row possess a higher number of neighboring cameras to collaborate with. Thereby, for instance in case of 25 cameras, an optimal use of resources is achieved and the system's performance is increased.

6.3 ACFSync (operation mode 1)

6.3.1 Experimental Setup

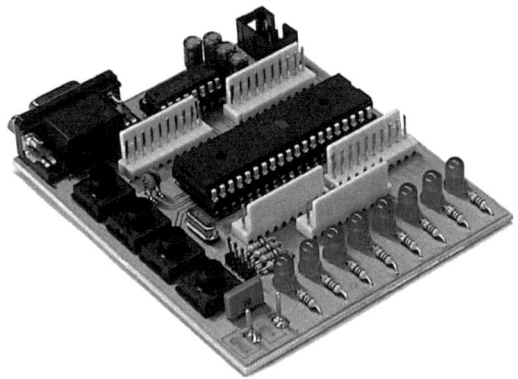

Figure 6.18: Photo of the experimental beacon sending a time stamp every 2 seconds [8]

The beacon used for testing purposes is depicted in Figure 6.18. The beacon contains a RS232 serial link, LEDs, an external $32.768\,kHz$ watch crystal, and a microcontroller. In normal operation, all LEDs are switched on and off synchronously in order to maximize the area of the signal source detectable by the camera.

6.3. ACFSYNC (OPERATION MODE 1)

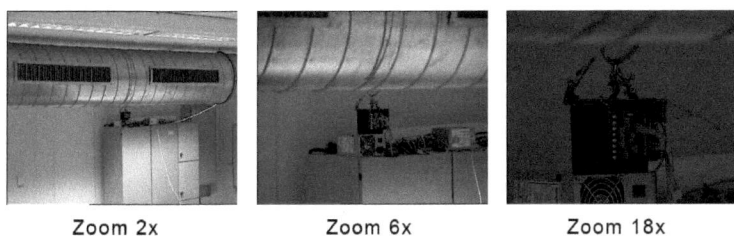

Zoom 2x Zoom 6x Zoom 18x

Figure 6.19: Zoom levels of the PTZ camera utilized to change the signal area

A Smart Camera, which consists an off-the-shelf laptop (Intel Pentium Atom, $1.6\,MHz$) attached to an Axis PTZ214 camera via Ethernet, is situated in a distance of approx. 6 meters to the beacon (with free field of view). Both, the beacon and the camera, are mounted at a height of approx. 2.5 meters. The Axis PTZ214 camera captures its environment with CIF resolution (352x288 pixels). The beacon sends an optical message containing a time stamp. The optical message is directly received by the camera, since it propagates with speed of light. As explained in Section 5.3.3, an optical message is sent every two seconds. Each optical bit has a length of $20\,ms$. After the image sensor of the IP camera has received the optical message, the image data is retrieved by the laptop via Ethernet, timestamped, and decoded in order to extract the time stamp. Afterwards, the laptop's local clock is set accordingly.

6.3.2 Synchronization Accuracy

In order to synchronize the local clock to the received time stamp, the client needs to be able to exactly determine that point in time – in terms of the local clock – of the first rising edge of the message received. Since this edge conforms to exactly one frame – or the time between two frames – captured by the camera, the necessary information is the exact time that passes between capturing a frame and the arrival of that frame in the part of the algorithm that manages this timing information.

138   CHAPTER 6. EVALUATION

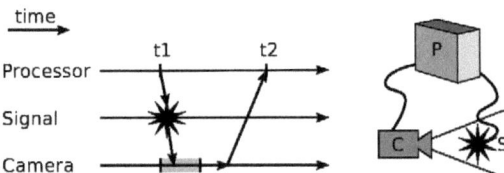

Figure 6.20: Experimental setup of a beacon and a Smart Camera capturing the beacon signal. A time stamp is sent by the beacon every 2 seconds.

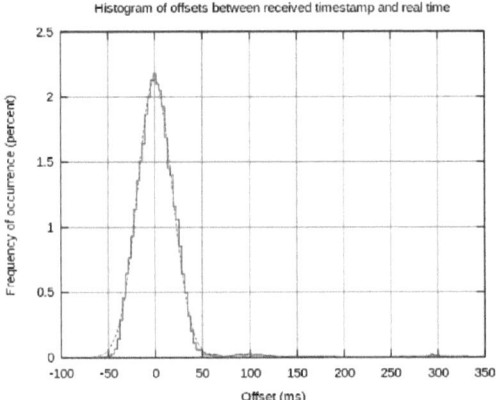

Figure 6.21: Histogram of offsets between received time stamp and wall clock

6.3. ACFSYNC (OPERATION MODE 1)

This delay between capturing and processing a frame ultimately depends on the nature of the link connecting the camera sensor (the IP camera) to the processing unit (the laptop). Consequently, the precise value needs to be measured – separately for each type of client – during some form of calibration phase. Such initialization may be unnecessary if the image capturing, the time keeping, and the image processing components are located in the same domain, one example is a single FPGA (field programmable gate array) due to the tighter coupling of the parts. Nevertheless, it is crucial to measure this delay in our setup, since it determines the synchronization accuracy.

The experiment as depicted in Figure 6.20 has been conducted to measure this delay. For this purpose, a camera is pointed at the beacon's source signal and a simple image processing algorithm has been implemented reacting on sharp changes in brightness. The camera as well as the beacon are connected to the same processing unit, whereas the beacon is connected to the serial port. One time stamp is taken at the time the signal is generated and another one is taken once the frame containing the change is detected. This process is repeated 100 times. The average time difference corresponds to the delay of reception ($140\,ms$). Figure 6.21 presents the frequency of occurring deviations. In 71 % of the cases, an offset of $\pm 20\,ms$ arises, i.e. $\pm$ the camera's frame rate. The offsets above $50\,ms$ stem from TCP/IP network latencies on the link between the IP camera and the laptop. Thus, a synchronization accuracy of $\pm 20\,ms$ can be achieved in 71 % of the cases, if the delay of reception is known.

6.3.3 Error Rate

The strength of the beacon signal is driven by the LED intensity and the zoom level of the PTZ camera. The zoom levels as illustrated by Figure 6.19 determine the amount of pixels in the imagery stemming from the beacon signal. The error rate measures how many optical messages can be decoded successfully, i.e. the fraction passing the CRC check. The experiment was conducted for two types of mask creation: (1) the algorithm had to find the optical beacon signal in the imagery on its own (see Figure 5.13), and (2) the algorithm was pointed to the beacon signal manually. In case of dynamic mask

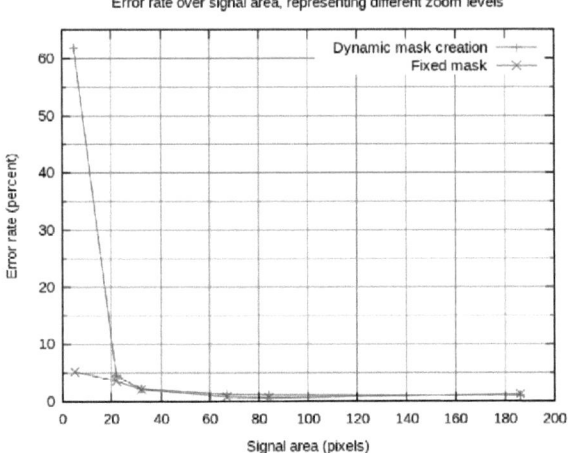

Figure 6.22: Error rate over signal area (represented by different zoom levels)

creation, the algorithm needs a beacon signal which has a size of at least 20 pixels in the captured image (see Figure 6.22), since a lower signal area leads to instabilities in terms of signal detection (e.g. due to thermal noise and low pass filtering). In case of a fixed mask, a beacon signal of 5 pixels is sufficient to achieve error rates below 5 %.

In addition to the signal area in pixels, the signal strength is driven by the LED intensity. The error rate has been measured for a fixed mask setting, i.e. the algorithm did not have to find the beacon signal on its own (see Section 5.3.2 for an explanation of dynamic mask creation). The camera captured the images in normal and over-exposed mode (see Figure 6.24) in order to test the sensitivity of the algorithm in terms of different lighting conditions. As depicted in Figure 6.23 for low LED intensities, i.e. below 10 % of the maximum LED intensity, the over-exposed mode performs better than the normal mode. The reason is that the contrast of the signal is improved by over-exposition. In terms of high LED intensities, i.e. higher than 60 %, over-

6.3. ACFSYNC (OPERATION MODE 1) 141

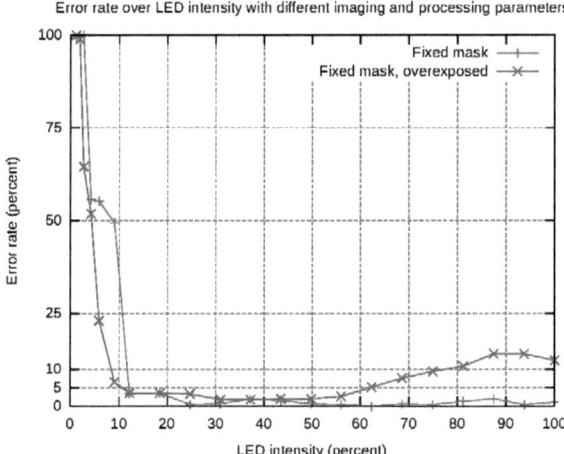

Figure 6.23: Error rate over LED intensity with fixed mask

exposition leads to an increased error rate due to non-linearities in terms of the image sensor. Therefore, the beacon signal (i.e. the camera's zoom level and the beacon's LED intensity) has to be adapted to the environmental conditions to achieve low error rates (i.e. the illumuniation and viewing distance).

6.3.4 Time Complexity

The time complexity is investigated in terms of the synchronization time. The synchronization time measures the time period between signal detection and successful decoding, i.e. the CRC check. This experiment has been conducted for varying LED intensities and different lighting conditions, which are emulated by changing the exposure time of the camera. The different exposure times are illustrated by Figure 6.24. The beacon sent optical messages constantly, whereas the Smart Camera captured 100 optical messages at points of time that were randomly distributed over a time period of 10 minutes. Thus,

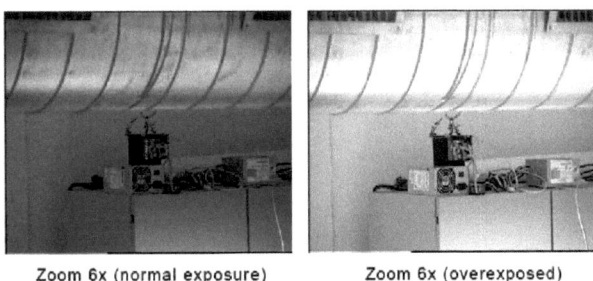

Zoom 6x (normal exposure)   Zoom 6x (overexposed)

Figure 6.24: Normal and over-exposed mode (for zoom level 6x)

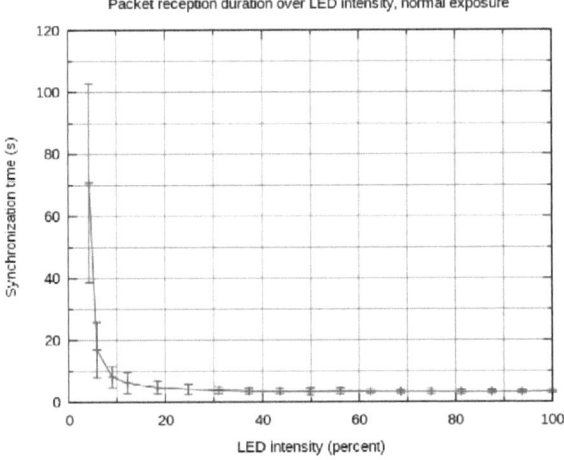

Figure 6.25: Packet reception duration over LED intensity, normal exposure

6.3. ACFSYNC (OPERATION MODE 1) 143

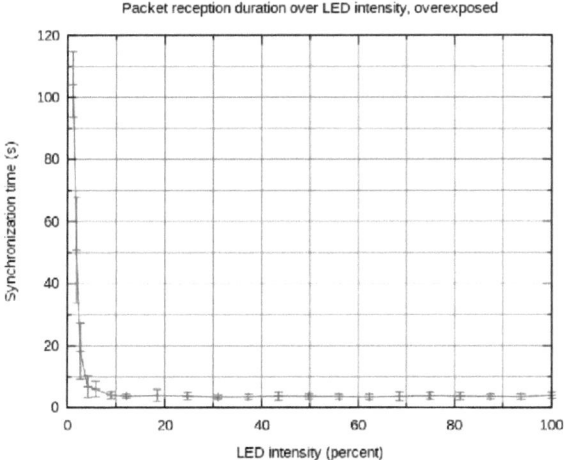

Figure 6.26: Packet reception duration over LED intensity, over-exposed

we emulated the situation of an Active Camera arriving at a beacon. As depicted in Figure 6.25, low LED intensities lead to high packet reception durations in normal exposure mode. This is caused by the high error rates which arise for low intensity values as explained in the previous section. For high LED intensities, the synchronization time decreases to $3.2\,s$ (the theoretical optimum is $3\,s$, assuming that the AC arrives in 50% of the cases too late, and too early, respectively), whereas the standard deviation decreases as well.

In case of over-exposition, a synchronization time of 3.2 seconds can already be achieved for LED intensities above 10% due to the improved signal contrast by over-exposition (see Figure 6.26). Nevertheless, over-exposition leads to non-linearities in terms of the image sensor and the mean synchronization time does not increase significantly for high LED intensies, only the standard deviation increases.

Figure 6.27: Person moving with 1.5 $\frac{m}{s}$ from right to left through a camera's field of view

6.3.5 CPU and Memory Utilization

The CPU and memory utilization have been measured on the Smart Camera for different resolutions and 600 seconds. In case of CIF resolution (352x288 pixels), the average CPU utilization is 36.09 % with a standard deviation of 3.57. In case of 2CIF resolution (704x288 pixels) the average CPU utilization increases to 63 %, whereas the standard deviation increases to 6.47. According to the amount of pixels, the CPU utilization is increased as well. The memory utilization is about 11 $MByte$ for CIF resolution and 15 $MByte$ for 2CIF resolution.

6.4 ACFSync (operation mode 2)

6.4.1 Experimental Setup

To evaluate the performance of $ACFSync$ in terms of operation mode 2, i.e. the cooperative frame synchronization mechanism, a 3-D scene was set up consisting of one or more persons in a surveillance area, see Figure 6.27. By using the 3-D software Blender [75], various virtual cameras can be positioned around a scene capturing synchronized video sequences from different perspectives. This is illustrated by Figure 6.30. The persons traversing the surveillance area

6.4. ACFSYNC (OPERATION MODE 2) 145

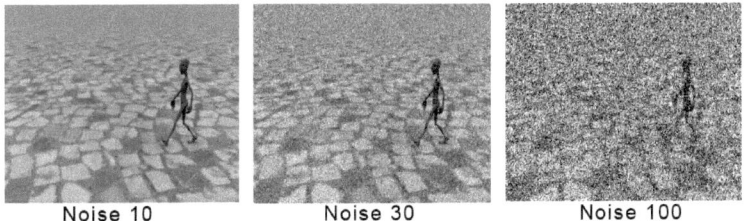

Figure 6.28: Different noise levels

are based on a 3-D model of a person, which is depicted in Figure 6.27. In case of scenarios consisting of one person, the 3-D person enters the surveillance area and moves with a speed of 1.5 $\frac{m}{s}$. After 2 seconds of moving, it stops to look at his watch. Hence, a salient behavior is generated.

Virtual cameras are positioned on a height of 6 meters above the scene and have free field of view on the scene. They capture their environment with CIF resolution, i.e. 352x288 pixels, at 25 frames per second. Each video sequence has a length of 180 frames, whereas the correlation horizon is set to 50 frames, i.e. 2 seconds. All videos are post-processed using a standard Gaussian noise (see Section 6.4.2) to emulate a minimal thermal noise of realistic cameras.

6.4.2 Noise

A virtual camera is positioned vertically to the moving person. It corresponds to the camera 0° as depicted in Figure 6.30. Different noise levels are added to the captured video sequence. Afterwards, autocorrelation is used for calculating the frame's offset and its certainty. The noise levels are presented in Figure 6.28. The noise is a Gaussian white noise that is added to every pixel value. The standard deviation is increased to a value of up to 100, which is added to each pixel's intensity value. The results are shown in Figure 6.29.

As illustrated by Figure 6.29, the calculated frame offset is zero for noises up to 20 and the certainty that this frame offset is correct is high. A noise higher than 30 leads to incorrect frame offsets. The noise interferes with

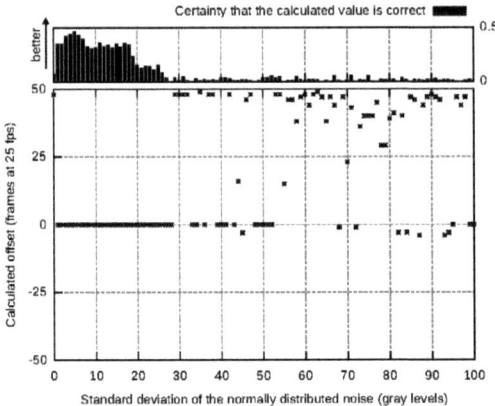

Figure 6.29: Frame offset for autocorrelation under the influence of additive normally distributed noise

the workflow of the underlying optical flow algorithm. Due to the noise, the optical flow algorithm is not able to find significant feature points any more. Original feature points are cropped as well as new artificial ones are added. Nevertheless, a noise level higher than 20 corresponds to a very strong noise (see Figure 6.28), which is usually not produced by thermal noise.

6.4.3 Perspective

In order to investigate the performance of the method towards different perspectives, three settings were chosen:

1. 36 cameras are aligned on a circle around the surveillance area and each video sequence is correlated to the video sequence of the 0° camera (see Figure 6.30)

2. 15 cameras are aligned on an arc over the surveillance area vertically to the animated walker's direction of movement (see Figure 6.32)

6.4. ACFSYNC (OPERATION MODE 2) 147

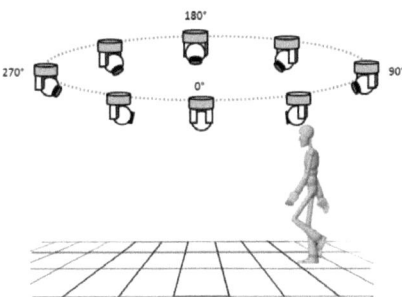

Figure 6.30: 36 cameras positioned each 10 degrees around the surveillance area

3. 15 cameras are positioned on an arc over the surveillance area horizontally to the animated walker's direction of movement (see Figure 6.34)

The results of aligning 36 cameras on a circle around the scene are depicted in Figure 6.31. A frame offset of zero and the highest certainty value are achieved for the auto-correlation of camera 0° with itself. The correct frame offset is achieved for a perspective of up to 45°. Afterwards, the frame offsets are between 25 and 50 frames. Due to the perspective, the saliency curve of the salient motion (looking at the watch) varies. Since looking at the watch lasts about 2 seconds (approx. 50 frames), the frame offset varies in this interval. Good results can be achieved by correlating video sequences of perspectives between 170° and 200° with 0°. In case of 180°, the maximal certainty is obtained, since it is the counterpart to the 0° camera. In summary, counterpart cameras, i.e. cameras which have an offset of 180° to each other, can be correlated with high accuracy and certainty. In addition, good results are achieved for correlating cameras having an offset of less than 45° to each other.

Since the best results are achieved for correlating the 0° camera with the 180° camera, we investigated whether the frame offset and certainty can be

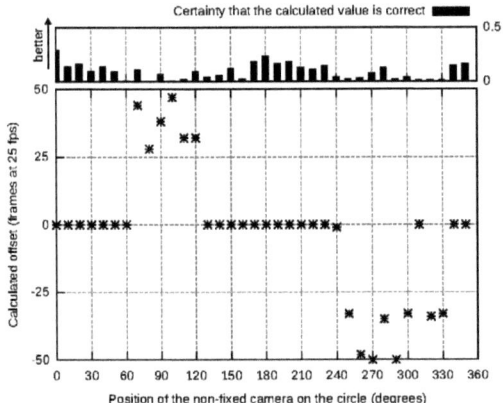

Figure 6.31: Calculated frame offset between the 0° camera and cameras 10° to 350° (on a circle around the surveillance area)

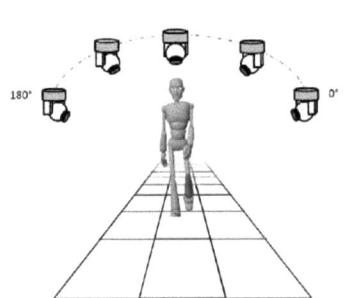

Figure 6.32: 15 cameras positioned on an arc over the surveillance area vertically to the direction of movement

6.4. ACFSYNC (OPERATION MODE 2)

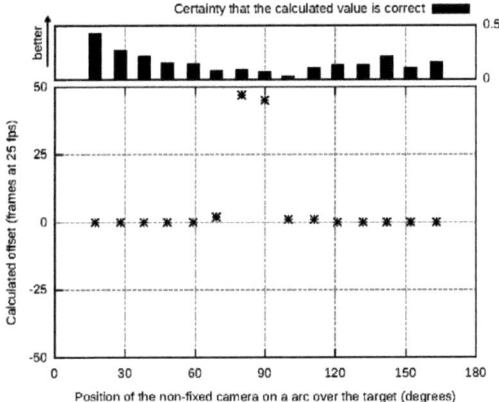

Figure 6.33: Calculated frame offset between the 0° camera and 15 cameras positioned on an arc vertically to the direction of movement

improved if the 180° camera is varied in terms of its z-axis' view angle. Therefore, 15 virtual cameras are positioned on an arc over the surveillance area as depicted in Figure 6.32, whereas a z-axis' view angle of 15° corresponds to the 0° camera in the circle setup and a z-axis' view angle of 175° to the 180° camera in the circle setup. The results are given in Figure 6.33. Since the 15° camera corresponds to the 0° camera in the circle setup, the best result is achieved for correlating this pair. Afterwards, the higher the z-axis' view angle is the worse is the correlation of the video sequences of both cameras. Thus, the best correlation results are achieved if the z-axis' view angle is increased up to 30°, since the person's motion is pronounced in a stronger way in such a setup.

In order to find the local minimum, 15 cameras have been aligned horizontally to the direction of movement, i.e. according to the 90° camera in the circle setup (see Figure 6.34). The z-axis' view angle is changed. A z-axis' view angle of 15° corresponds to the view angle of the 90° camera in the circle setup, whereas an z-axis' view angle of 175° corresponds to 180° in the circle

150   CHAPTER 6. EVALUATION

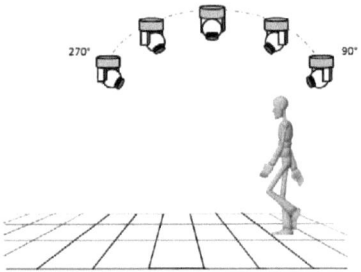

Figure 6.34: 15 cameras positioned on an arc over the surveillance area horizontally to the direction of movement

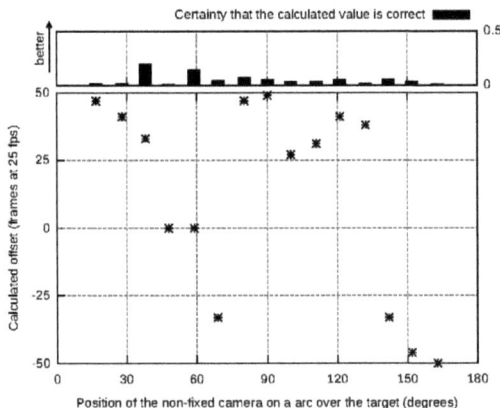

Figure 6.35: Calculated frame offset between the 0° camera and 15 cameras positioned on an arc horizontally to the direction of movement

setup. As depicted in Figure 6.35, there is no view angle along this arc improving the correlation results, since a view angle of 90° in the circle setup does not pronounce the person's movement in a strong way. Thus, the optical flow algorithm has weak motion vectors.

6.4.4 Number of Targets

In order to investigate the robustness of the method in terms of the number of targets in the surveillance area, two scenes have been set up:

1. A scene with two persons (pair scene)

2. A scene with seven persons (crowded scene)

Nevertheless, only one target has the salient behavior, i.e. looking at his watch. The rest of the persons only traverse the scene. As depicted in Figures 6.36 and 6.37, increasing the number of targets within in surveillance area significantly reduces the quality of the correlation results. In the pair scene (see Figure 6.36), our method still achieves correct frame offsets for correlating the 0° camera with its counterpart camera, but the certainty that this correlation is correct is decreased. Increasing the number of targets to seven leads to unacceptable correlation results, since the salient behavior is disturbed by the motion vectors of the other targets within the scene.

Nevertheless, in order to avoid this situation in real-world scenarios, mechanisms could be used to detect time intervals with crowded scenes [76] and avoid a resynchronization during these periods. During time periods with reduced activity, the aforementioned synchronization method could be started yielding good synchronization accuracies.

6.4.5 Real-world Experiment

In order to evaluate the real-world applicability of our method, we conducted the following experiment: We performed experiments with two *PTZ Axis 214*

152   CHAPTER 6. EVALUATION

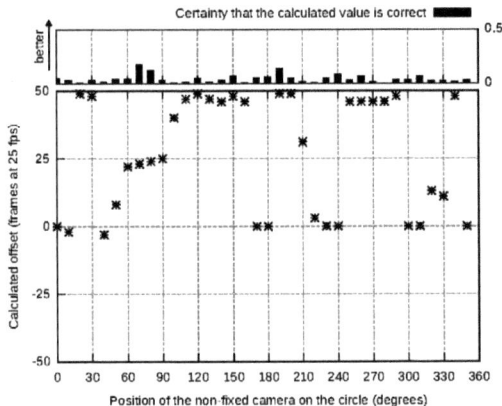

Figure 6.36: Calculated frame offset between the 0° camera and cameras 10° to 350° on a circle around the surveillance area with two persons (pair scene)

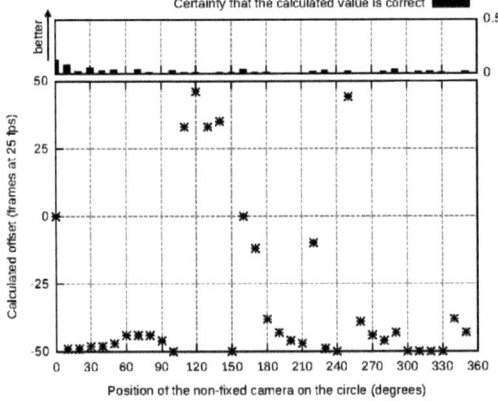

Figure 6.37: Calculated frame offset between the 0° camera and cameras 10° to 350° on a circle around the surveillance area with 7 persons (crowded scene)

6.4. ACFSYNC (OPERATION MODE 2)

IP cams capturing pictures (QCIF resolution, 176x144 pixels) of an office hallway. The IP cams were situated in an office with an open door. Hence, they have a free field of view on the hallway. The cameras were situated on an office table with a distance of 25 cm to each other. Their axis of view were parallelized and they possessed overlapping fields of view without any particular calibration. Thus, the difference of their view angles is less than 45°, which leads to good correlation results as described in Section 6.4.3. Every IP cam was connected via the university's WLAN (IEEE 802.11g) to a laptop (Intel Pentium Atom, $1.6\,MHz$) (the building is covered by the university's WLAN access points (APs) using channel 11 on 802.11g). Besides these APs, the environment contains microwave ovens, wireless sensors, and Bluetooth devices that may potentially interfere with the wireless transmissions of our IP cams. The laptops retrieve the JPEG-compressed pictures from the IP cams via HTTP and perform our frame synchronization algorithm. Both devices are synchronized via NTP (NTP server: time1.rrzn.uni-hannover.de). We expect that such a rather "chaotic" wireless environment is typical for a real-world deployment of embedded Smart Cameras (given increased popularity of home APs, wireless sensors, and community meshes).

Pictures are captured with a frame rate of $25\,fps$. After having observed 100 events (especially persons crossing the cameras' field of view on their way to the kitchen), these events were used to measure the method's accuracy. The video streams of both cameras were evaluated in terms of their saliency curves and the frame synchronization method was executed on this data. One result is that video streams with events of short duration ($< 10\,frames$) occur very seldom, see Figure 6.38, but lead to high synchronization accuracies (i.e. $2.5\,frames$, standard deviation of $2.12\,frames$). Events of long duration ($> 10\,frames$), see Figure 6.39, occur more often but lead to high synchronization errors. The mean synchronization error is $6.62\,frames$ with a standard deviation of $2.74\,frames$. If we consider events with a maximum length of 10 frames only, the mean error is about $2.5\,frames$ with a deviation of $2.12\,frames$. To avoid inaccuracies due to long events, they could be filtered online on the Smart Camera, by only considering events with significant peaks below the length of 10 frames. Another solution can be to configure the cam-

154   CHAPTER 6. EVALUATION

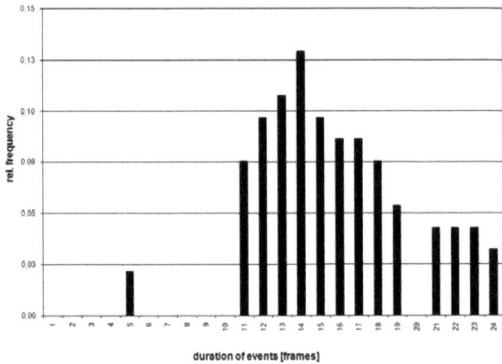

Figure 6.38: Relation of event duration and frequency of occurrence in an office hallway scenario

eras' field of view in a way that events can only occur in short time periods, e.g. by using the camera's zoom functionality.

6.4.6 CPU and Memory Utilization

Our experiments for measuring the CPU and memory utilization were carried out in the same setup as described in the previous section. We captured pictures with a frame rate of 25 fps for 20 minutes. Every minute, a correlation computation was executed on the device to emulate the appearance of a salient event in the video data stream, which is to be synchronized with the video data stream of another device. The mean length of these events is approx. 50 $frames$. The data of the other's device saliency curve is loaded via a SOAP interface. The CPU utilization varies between 50 % and 60 %. The mean CPU utilization is about 54.99 %. Although this is a high CPU utilization, future Smart Cameras will be equipped by more powerful processors. Thus the CPU utilization is assumed to decrease on future systems. Nevertheless, this will always account for a basic CPU utilization of approx. 50 % to 60 %, since

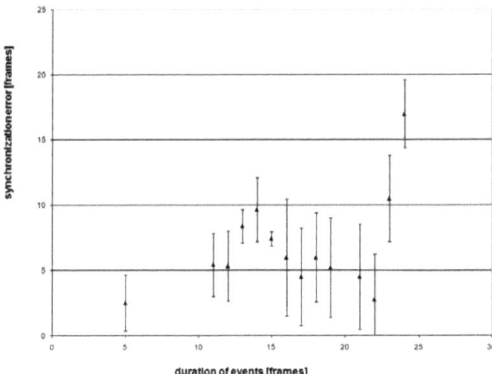

Figure 6.39: Average synchronization error with standard deviation in an office hallway scenario

our application scenario is dominated by surveillance tasks (i.e. capturing pictures and image processing like object detection or tracking) constantly. The memory utilization is about 20 $MByte$. As today's Smart Cameras are equipped with at least 1 GByte of RAM and future systems are assumed to possess more RAM, this is a reasonable value.

6.5 Summary

In this chapter, we have presented an extensive evaluation of both dynamic reconfiguration methods, the *DRofACN* and *ACFSync* method, which are described in Chapter 5. For the experiments, the important system parameters of *DRofACN* (e.g. system size and induced load) as well as important parameters of *ACFSync* (e.g. view angle and the beacon's signal strength) have been varied. In this section, we present a summary of the results and give recommendations for ideal conditions of both methods.

6.5.1 DRofACN

In Section 6.2.2, we analyzed the scalability in terms of different target generation rates and system sizes and showed that an Active Camera Network of 100 nodes can handle up to 2,500 targets of interest simultaneously with a TAR ratio of 90 % and a mean target detection of less than 10 seconds. In order to increase the method's performance, i.e. the TAR ratio or the mean target detection time, the number of resources has to be increased within the network. A TAR ratio of up to 80 % can be achieved by increasing the number of cameras in the network according to the target generation rate. In order to achieve a TAR ratio higher than 80 %, the number of resources needed for a specific target generation rate does not increase linearly any more. Here, significantly more resources are needed.

In Section 6.2.3, we showed that packet loss concerning the camera-to-camera communication has minor impact on the system's performance, i.e. the TAR ratio. For scenarios with up to 9 cameras, the TAR ratio is only decreased by 10 %. Nevertheless, if the number of ACs is increased, packet loss has a higher impact on the TAR ratio. For example, for a network consisting of 36 ACs, TAR is decreased by approx. 20 %. The reason is that networks which consist of a higher number of cameras are affected in a stronger way, since packet loss has a negative impact on the coordination among the nodes.

In Section 6.2.4, we investigated how localization errors of targets stemming from perceiver nodes influence *DRofACN*'s performance. Target localization errors up to $20\,cm$ can be tolerated by the system. The more resources are available the less is the influence of localization errors on the system's performance. Localization errors up to $20\,cm$ can arise in perceiver networks consisting of LASER scanners.

In Section 6.2.5, we analyzed *DRofACN*'s performance, if the speed of targets is increased. For low values ($< 5\,\frac{m}{s}$), the load induced by the increased target speed can be compensated by increasing the number of cameras within the network. If the target speed exceeds a specific value ($8\,\frac{m}{s}$), this effect does not hold any more. This behavior is associated with the sensing range of the perceiver nodes connected to the cameras. By increasing the sensing range,

6.5. SUMMARY

scalability can also be achieved for high target speeds.

In Section 6.2.6, we showed that mechanisms for phenomenon adaptivity can be used to increase *DRofACN*'s performance. We conducted experiments for different node placements and spreads and showed that the TAR ratio can be improved significantly in case of trajectories with small spreads (varied between $2.5\,m$ and $250\,m$).

In summary, it can be concluded that using *DRofACN* is attractive, if a TAR ratio of less than 80% and a mean target detection time of less than 20 seconds is sufficient. In order to achieve a TAR ratio higher than 80%, the number of resources does not increase linearly any more. Additionally, in systems that are exposed to high packet loss concerning the communication channel or localization errors (in terms of the targets), it showed to be very robust.

6.5.2 ACFSync

In Section 6.3, we showed that concerning the beacon-based synchronization method (*ACFSync* operation mode 1) a synchronization accuracy of $\pm 20\,ms$, i.e. $\pm$ the camera's frame rate, can been achieved in 71% of the cases. Additionally, we anaylzed the error rate, i.e. of wrong decoding of the optical messages, in case of dynamic mask creation and fixed mask creation. The algorithm needs a beacon signal which has a size of at least 20 pixels in the image captured, since a lower signal area leads to instabilities in terms of signal detection. In case of a fixed mask, a beacon signal of 5 pixels is sufficient to achieve error rates below 5%.

In Section 6.4, we presented evaluations for the cooperative frame synchronization method based on visual events (*ACFSync* operation mode 2). This method achieves good correlation results for counterpart cameras, i.e. cameras which have a viewing offset of 180° to each other. Good results can also be achieved for correlating video sequences of cameras having a viewing offset of less than 45° to each other. Nevertheless, increasing the number of targets within the surveillance area leads to unacceptable correlation results. The reason is that the salient behavior is disturbed by the motion vectors of the other targets within the scene. Furthermore, we investigated the real-world applicability of the algorithm. In terms of short visual events (i.e. below 10 frames), good synchronization accuracies can be achieved. Nevertheless, long events (i.e. above 10 frames) lead to inaccuracies. To avoid inaccuracies due to long events, they can be filtered online on the Smart Camera, by only considering events with significant peaks below the length of 10 frames. Another solution can be to configure the cameras' field of view in a way that events can only occur in short time periods, e.g. by using the camera's zoom functionality.

In summary, it can be concluded that using *ACFSync* is attractive in non-crowded scenes. This is important to detect and decode the beacon-signal correctly as well as for the correlation algorithm of the frame synchronization method. Nevertheless, in non-crowded scenarios frame accuracy can be achieved.

Chapter 7

Related Work

First, we review the state of the art of research fields related to the problem of dynamic reconfiguration in Section 7.1. In the past, dynamic reconfiguration has been used in real-time systems as well as for planning issues in the realm of vehicle routing and sensor planning.

Secondly, we review the state of the art relevant for our system architecture as introduced in Chapter 4. For this purpose, we discuss middlewares for embedded and organic systems in Section 7.2, since they are most related to our system architecture for Active Cameras.

Afterwards, in Section 7.3 and 7.4, related work in the field of Active Cameras and Active Vision as well as time synchronization is presented and discussed in relation to the dynamic reconfiguration methods we developed in Chapter 5.

Finally, we conclude with a summary.

7.1 Dynamic Reconfiguration

In this section, we present a broad overview about research fields related to the general problem of dynamic reconfiguration.

7.1.1 Scheduling

Scheduling is the process of deciding how to commit resources between a variety of possible tasks. Time can be specified (with hard time constraints) or floating as part of a sequence of events. Therefore, we review state of the art of scheduling mechanisms in (non) real-time systems in the following two subsections.

Real-time Systems

Modern real-time systems are designed for adaptivity by using dynamic reconfiguration. This reconfiguration allows to react to aperiodic events in a predictable manner. Thus, graceful degradation in overload scenarios can be guaranteed whenever needed. In this context, they are structured as a set of multiversion tasks in order to implement services with various levels of quality. In overloaded scenarios, for instance, a lower quality service may be scheduled for execution keeping the system's correctness and providing graceful degradation. The goal of the reconfiguration mechanisms is to select the versions of tasks that lead to the maximum benefit for the system at runtime.

A very simple strategy for dynamic reconfiguration of real-time systems is temporal protection. Here, admission control mechanisms reject or cancel tasks if needed [77, 78]. This approach is used for soft real-time applications due to their tolerance to missing deadlines. Nevertheless, this is not appropriate for all classes of real-time systems due to this behavior.

The solution proposed by Jehuda and Israeli [79] also deals with multiple versions of tasks. The solution aims at maximizing the system benefit subject to the schedulability conditions, which are expressed as processor utilization bounds. The reconfiguration problem is modeled as the classical knapsack problem, where the boundaries of the processor utilization represent the knapsack's size. Nevertheless, this approach is not able to deal with aperiodic tasks, which is necessary for dynamic environments.

Rusu et al. [80] proposed a reconfiguration mechanism for energy-aware real-time systems. The system benefit is optimized considering both schedulability and energy constraints. The authors provide reconfiguration at the

7.1. DYNAMIC RECONFIGURATION

scheduler level. Nevertheless, all tasks are required to have the same periods and deadlines, which is not an acceptable assumption for dynamic environments.

Lima et al. [81] propose a reconfiguration mechanism, which selects appropriate task versions that maximize the global benefit for the system. The system schedulability conditions are evaluated at runtime by means of dynamic programming techniques. Nevertheless, the reconfiguration mechanism runs a single scheduler. Thus, the mechanism does not scale in contrast to our method.

Non Real-time Systems

Scheduling is not only used in real-time systems. Load balancing can be interpreted as a specific form of scheduling, e.g. for a parallel system it is one of the most important problems which has to be solved in order to enable the efficient use of parallel computer systems. Load balancing is a computer networking methodology to distribute workload across multiple resources, e.g. computers, computer cluster or other resources. The goal is to achieve optimal resource utilization, maximize throughput, minimize response time, and avoid overload. Using multiple components with load balancing, instead of a single component, may increase reliability through redundancy.

A broad overview can be found in [82]. Nevertheless, Active Camera Networks come with specific requirements in terms of computer vision and time constraints, e.g. to allow for image processing tasks, which are not fulfilled by traditional load-balancing mechanisms.

7.1.2 Dynamic Vehicle Routing Problem with Time Windows

The Dynamic Vehicle Routing Problem with Time Windows (DVRPTW) addresses the problem of dynamic reconfiguration [83]. The Vehicle Routing Problem (VRP) is a combinatorial optimization problem in distribution logistics. A VRP aims to get the best vehicle routing and scheduling for one or several vehicles driving from a depot to customers and back to the depot without exceeding the vehicles' capacity constraints. The Dynamic Vehicle Routing Problem (DVRP) has drawn more attention in the recent past. It allows vehicles to update services based on renewed information [84]. The input parameters are predefined and static in typical VRPs. Allowing the change of parameters, e.g. customer demands, during runtime leads to the DVRP. It is also called the real-time, or the on-line problem [83].

In order to solve the DVRP mainly two strategies are followed:

1. Adapt algorithms from the typical VRP domain to trigger the re-optimization periodically

2. Use of stochastic methods

In terms of adapting algorithms from the typical VRP domain, several approaches have been proposed, e.g. [85, 86]. They are based on standard heuristics such as simulated annealing, tabu search, genetic algorithms or ant colony optimization and use wrapping mechanisms in order to trigger re-optimization from time to time. The second strategy is based on stochastic methods, such as the Markov Decision Process and Stochastic Programming [87]. However, these approaches are limited in handling large-scale dynamic scenarios.

The Vehicle Routing Problem with Time Windows (VRPTW) is an extension of the classical VRP and is defined as follows: Given a set of depots, a homogeneous fleet of vehicles and a set of known demand locations, find a set of closed routes, originating and ending at the depots, that service all demands and minimize the travel cost; in addition, the service at each demand location must start within an associated time window. All problem parameters, such as demand locations and time windows, are assumed to be known with certainty. Time window's constraints are indeed common in many applications, including bank deliveries, postal deliveries, grocery distribution, dial-a-ride service, bus routing, and repairmen scheduling. The VRPTW has generated significant research interest over the years [88, 89, 90], resulting in major contributions in the area of combinatorial optimization. However, many scenarios in the realm of the VRPTW are not static and deterministic such as routing problems [91].

7.1. DYNAMIC RECONFIGURATION

In fact, requests for service often arrive sequentially in time, and these arrival epochs may be stochastic. Additionally, locations of future demands may be unknown or known only probabilistically.

Due to this reason, researchers combined both problems and created the Dynamic Vehicle Routing Problem with Time Windows (DVRPTW). E.g. in [92], the problem of designing motion strategies for a team of mobile agents is studied, in order to fulfill requests for on-site service in a given planar region. Each service request is generated by a spatio-temporal stochastic process and remains active for a certain deterministic amount of time, and then expires. An active service request is fulfilled when one of the mobile agents visits the location of the request. Nevertheless and in contrast to our work, they assume the on-site service time to be zero. In our work, the on-site service time is assumed to be non-zero for applicability of computer vision algorithms. In addition, we assume the expiration time of the target to be dynamic, since it depends on the velocity of the targets moving through the workspace.

7.1.3 Sensor Planning for Visual Surveillance

Traditionally, dynamic reconfiguration has also been utilized for active sensor planning to determine the configuration of a set of sensors in static or dynamic surveillance environments. Therefore, we also review the state of the art and recent developments in this research field.

The primary focus of the sensor planning domain was focused on the analysis of placement constraints in static environments. These constraints were mainly resolution, focus, field of view, visibility, and conditions for light source placement in 2-D space [93]. This was driven by the requirement to place a viewpoint in an acceptable space, whereas several prerequisites should be satisfied. In [94] several approaches are presented in terms of sensing strategies, which were developed between 1987 and 1991. Among these approaches, Cowan et al. [95] investigated the computation of acceptable viewpoints for satisfying optical requirements concerning the sensor, i.e. sensor placement constraints. E.g. the lens aperture setting or light position region were determined in order to achieve adequate illumination conditions. Abrams et al.

[96] proposed to compute the viewpoints in relation to other optical constraints such as resolution, focus (depth of field), field of view, and detectability. In summary, early work focused on optical constraints rather than coordination issues. In contrast to our method, a static environment was assumed.

Recently, there has been greater interest in sensor planning in dynamic environments [97, 98], especially for tracking applications. Nevertheless, the majority of systems utilize methods developed for sensor planning in static environments. E.g. the system presented in [97] is based on an off-line heuristic, which computes sensor motions in 2-D fed by an *a priori* known object trajectory. The on-line controller of the system is responsible for the re-adjustment of the sensor motions in case of deviations of the actual object's trajectory from the expected one. In contrast, a system is presented in [99] that does not require any *a priori* knowledge about the target's trajectory in order to fulfill the sensor-planning task. The system splits up the workspace in discrete sectors and in case an object enters a sector, sensors responsible for the surveillance of this sector provide information about the object. In [100], a system is proposed, which is based on multiple sensors. These sensors are able to determine their path independently through a triangulation method. Thus, they are able to avoid obstacles. The system described in [101] aims at optimizing the amount of the targets that can be observed at any given time. This is accomplished through a coordination mechanism, which is based on negotiation techniques. In [102], a system is presented, which handles both, sensor placement with constraints and the pose estimation of the target, by using a Bayesian network for task-specific sensor planning. The Bayesian network is reconstructed at a constant rate in order to consider dynamics concerning the target's pose and position. Based on this, the active sensors are repositioned in a way that the target visibility is maximized and the sensing cost (i.e. movement of the sensor) is minimized. In [1], a method is presented for selecting and positioning groups of sensors in a coordinated manner for the surveillance of a maneuvering object. The object's trajectory is estimated on the basis of historical data and divided into time slots. These time slots are assigned to a sub-group of sensors, which are repositioned in order to observe the object. No *a priori* information is needed. Nevertheless, in case *a priori* information is available, the system is able to calculate initial sensor locations and orientations to increase the system's performance. In [103], a system is reported for sensor planning, which is used to compute the optimal positions for inspection tasks using known imaging sensors and feature-based object models. The initial setting is generated off-line, and on-line plans are computed for more complex tasks, called *inspection scripts*. Viewpoint optimality is defined as a function of feature visibility and measurement reliability. In [104], a methodology based on the notion of *attention-based behavior* is presented. Attention-based behavior systems rely on a supervisor to dynamically reconfigure the system, e.g. by selecting a single target for all the sensors to focus on for a certain period of time. Thus, a multi-target problem can be reduced to a single-target one. For instance, the system presented in [105] is based on a fuzzy controller to dynamically select targets on the basis of a set of expert rules. Afterwards, the orientations and settings of the cameras are reconfigured to optimize the resolution and the visibility of the targets.

In contrast to our method, none of the aforementioned systems uses reconfiguration in order to maximize the cameras' imaging quality. This means that their sensing performance metric does not take the requirements of the underlying computer vision algorithm into account. Moreover, our system offers dynamic reconfiguration for clock synchronization, since clock drift is an important driver for errors in terms of data fusion in surveillance systems.

7.2 Operating System and Middleware

In this section, we review the state of the art of middleware implementations. Section 7.2.1 gives background information on general-purpose middlewares. Secondly, middleware implementations for embedded systems are introduced in Section 7.2.2. Finally, we close with middleware implementations for organic systems in Section 7.2.3.

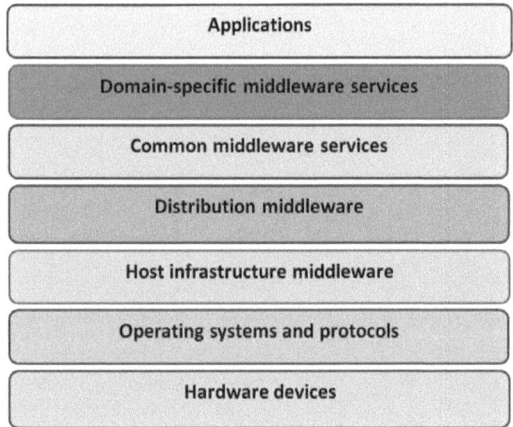

Figure 7.1: General-purpose middleware layers [9]

7.2.1 General-Purpose Middleware

Usually, middleware systems have to run on different hardware platforms and support various communication channels and protocols. In addition, they must be able to bridge applications running on different platforms, possibly in different programming languages, into a common distributed system. Due to this reason, a layered architecture is often used in order to support software flexibility on different levels. A very general partitioning into different layers of abstraction is depicted in Figure 7.1 as introduced in [9].

The *operating system* along with its hardware drivers, concurrency mechanisms, and communication channels is the basis of each middleware. In this layer, drivers for the underlying hardware platform are encapsulated. Furthermore, basic mechanisms for device access, concurrency, process and thread management are provided. The *host infrastructure layer* contains the low-level system calls in reusable modules. It also hides non-portable aspects of the operating system and is the first step toward a portable and platform-independent middleware. The *distribution layer* integrates multiple network hosts into a distributed system and defines higher-level models for distributed

7.2. OPERATING SYSTEM AND MIDDLEWARE

programming. The *common middleware service* layer augments the underlying distribution layer by defining domain-independent components and services which can be reused in applications and, thereby, simplify development. In terms of distributed applications, this could contain, for example, logging and global resource management. The *domain specific* layer provides services to applications of a particular domain in order to simplify their development. The highest level of this architecture is the *application layer*, where individual applications for a distributed system are implemented using services provided by the lower layers.

In terms of general-purpose middlewares, various middleware implementations have evolved during the last decades picking up some of the aforementioned layers. Probably the most prominent middleware standard is OMG's Common Object Request Broker Architecture (CORBA) [106]. CORBA is a distributed object system that allows objects on different hosts to interoperate across the network. In addition to CORBA, Real-Time CORBA (RT-CORBA) and Minimum CORBA haven been specified for resource-constrained real-time systems [107, 108]. Another middleware for networked systems is Microsoft's Distributed Component Object Model (DCOM) [109]. It runs on Window platforms only and allows communication of software components over a network via remote method invocation. Java Remote Method Invocation (RMI) [110], promoted by Sun, follows a similar approach. RMI allows invocation of an object method in a different Java virtual machine, possibly on a different host. Hereby, the development of distributed Java applications is simplified.

Nevertheless, Active Camera Networks come with specific requirements in terms of a system architecture, e.g. to allow for image processing tasks, which are not fulfilled by general-purpose middlewares.

7.2.2 Middleware for Embedded Systems

The requirements for middlewares for embedded systems, e.g. wireless sensor networks, are significantly different compared to those in general-purpose computing systems. These middleware systems usually focus on reliable services for ad-hoc networks and energy awareness [111], since wireless sensor networks are an inherently distributed system where individual sensors have to collaborate. The resources and capabilities of the nodes are very limited. In [112], Molla et al. have surveyed recent research on middlewares for wireless sensor networks. A result was that most implementations are based on TinyOS [113], a component-oriented, event-driven operating system for sensor nodes (motes). Several interesting approaches have been implemented and evaluated. The spectrum ranges from a virtual machine on top of TinyOS – in order to hide platform and operating system details – to more data-centric approaches for data aggregation.

Nevertheless, middlewares for embedded systems are not intended to cope with advanced image processing tasks and sending large amount of data as it is required by Active Camera Networks.

7.2.3 Middleware for Organic Systems

To allow for flexibility and adaptivity in terms of dynamic environments, various middlewares for organic systems have been presented. Furthermore, several of them might be suited for Active Camera Networks. To name a few, BASE [114], ORCA [115], and OCμ (former AMUN) [116] have been developed in this context. They offer many of those functionalities which are needed for Active Camera Networks. Nevertheless, they still lack some important points. For instance, OCμ offers a vast variety of functions which allow for designing Organic Computing systems. E.g. in [117], OCμ is used as integration platform of an artificial immune system. Other middleware architectures, such as BASE, offer a component-based approach. A system based on the BASE architecture could be designed according to the unit construction principle, which suits the needs of Active Camera Networks. In addition, the ORCA middleware delivers components for the control of mobile entities in a decentralized manner. Hoffmann [118] proposes a Smart Camera middleware allowing for self-organization in terms of spatial partitioning. Nevertheless, mobile Smart Cameras, dynamic environments as well as position control for improved image acquisition are not considered.

In general, it remains unclear how well existing middleware approaches can cope with the demands arising in Active Camera Networks. Especially, real-time capabilities are still subject to ongoing research. Another important aspect of a middleware for Active Camera Networks is the close coupling of all algorithms to the image data that is acquired by the image sensor. This data has to be analyzed in real-time and needs to be accessible from different components throughout the middleware. Apart from detecting events of interest, several system components can benefit from context information. Due to this reason, an image data centric approach seems to be most appropriate, which has been chosen in the context of this thesis. In addition, a new middleware has been designed focusing on the dynamic reconfiguration of Active Cameras and hence allowing for mobile scenarios. Thereby, in contrast to the aforementioned approaches, a lightweight and highly specialized middleware has become available, that exactly suits the needs arising in Active Camera Networks.

7.3 Active Cameras and Active Vision

Active Camera Networks are an expanding field of research. An extensive overview of current and past research is provided by Aghajan and Cavallaro [119]. First, we present related work in the realm of active cameras. This intends to show how the computer vision community utilizes activity of cameras to solve the placement problem. Secondly, related work in the field of active vision agents is presented, which focuses on the use of mobile robots for advanced target tracking.

7.3.1 Optimal Placement

The problem of finding optimal placement of stationary cameras within an observation area has long been studied. Once mounted in place, these cameras have a fixed position, orientation, and focal length. The earliest examination can be traced back to the "art gallery problem" in computational geometry. The problem is concerned with the question of how to place cameras in an arbitrary-shape polygon so as to cover the entire area [120]. Chvátal proved that the upper bound of the number of cameras is $\lfloor n/3 \rfloor$ [121]. Nevertheless, determining the minimum number of cameras turns out to be an NP-complete problem [122]. The problem of camera placement transforms to camera selection by incorporating additional constraints such as sensing range or priority of observation areas into the problem. Thus, the problem is reduced to the well-known set-cover problem [123]. Such a set-cover problem is again NP-hard. In the following, we present approaches for camera placement and selection.

The theoretical difficulties of camera placement are well understood, and many approximate solutions have been proposed. Nevertheless, few of them can be directly applied to real-world scenarios, since they are based on *a priori* knowledge and cannot guarantee the coverage of a predefined space inside a specific area with a minimum level of imaging quality such as image resolution, which is crucial for computer vision algorithms. Camera placement has been also studied in the field of photogrammetry for building accurate 3-D models. Various metrics such as the visual hull [124] and the viewpoint entropy [125] have been developed, and optimizations are performed by distinct types of ad-hoc searching and heuristics [126].

Based on these metrics, sophisticated modeling schemes for camera placement and selection were developed. Nevertheless, the sophistication of their visibility models comes at a high computational cost for the optimization. For instance, the simulated annealing scheme used by Mittal et al. [127] lasts several hours to find the optimal placement of four cameras in a room. Other optimization schemes such as hill climbing [128], semidefinite programming [129], and evolutionary approaches [130] are computationally intensive and prone to local minima.

Alternatively, the optimization can be performed in the discrete domain. Hörster et al. [131], for example, developed a flexible camera placement model by discretizing the space into a grid and denoting the possible placements of the camera as a binary variable over each grid point. Thus, integer linear programming can be used to compute the optimal camera configuration. Different constraints and cost functions can be integrated. Similar approaches have been published in [132, 133].

7.3. ACTIVE CAMERAS AND ACTIVE VISION

All of these techniques are based on assuming a very dense placement of cameras. Therefore, they are not applicable to real-world scenarios, which usually suffer from a sparse placement of cameras due to wide-areas or financial reasons. Nevertheless, these algorithms can be seen as the first configuration algorithms for Active Camera Networks, since they are able to compute an initial starting position for the system, although this is based on *a priori* knowledge and its quality decreases over time.

7.3.2 Active Cameras

In case of stationary cameras, a very dense placement is necessary to achieve the required image resolution for the underlying computer vision algorithms. Therefore, active pan/tilt/zoom cameras have been used to make way for dynamic camera selection at runtime. These cameras can rotate around their horizontal (tilt) and vertical (pan) axis using servos. Some of them are also equipped with an adjustable focal length (zoom) limited by a certain range.

A significant number of work in the literature has focused on object detection and tracking [134, 135, 136, 137]. Especially multi-target, multi-camera tracking in the context of multi-camera surveillance systems was investigated [138, 139, 140]. Accurate detection and tracking is crucial for these systems, since the extracted tracking information provides a basis for event detection [141, 142]. Additionally, tracking data is required in order to control a set of active cameras to acquire high-quality imagery [143] and object association within the system, e.g. through biometric signatures as presented in [144]. Ram et al. [145] proposed a framework to study the performance of visual coverage in wide-area scenarios, which unlike previous techniques, takes the orientation of the object into account. They defined a metric to compute the probability of observing an object of random orientation from one sensor and used it to recursively compute the performance of multiple sensors.

The integration of active cameras into stationary camera networks is usually based on a setup termed *master-slave* [146]. For this purpose, fixed and PTZ cameras are combined, which are calibrated beforehand by a human using calibration marks. Many researchers use a master-slave camera configuration with two [147, 148] or more cameras [149, 150] and assign targets to PTZ cameras based on scheduling algorithms. These heuristic-based algorithms provide a simple and tractable way of computing schedules. Nevertheless, they quickly become non applicable as the number of targets increases and exceeds the number of PTZ cameras, in which case the scheduling problem becomes increasingly nontrivial. A lack of PTZ resources must be considered when designing a scheduling strategy that aims at maximizing the number of captures per target and capturing as many targets as possible [94]. To solve this problem, researchers like Hampapur et al. [151] proposed a number of different camera-scheduling algorithms designed for various application goals. As one example, a round-robin method is described that assigns cameras to targets sequentially and periodically in order to achieve uniform coverage. Qureshi et al. [152] proposed an approach, where priorities are assigned to the targets by ordering them in a priority queue, e.g. depending on their arrival time. Bimbo et al. [153] ordered the targets according to the estimated deadlines by which they leave the area of observation. An optimal subset of targets, which satisfies the deadline constraint, is obtained through an exhaustive search. In the recent past, Li et al. proposed an dynamic camera assignment method using game theory [154]. Various criteria characterizing the performance of PTZ camera imaging and target tracking are presented in order to define utility functions. These functions are optimized through a bargaining mechanism in a game. The scheduling problem can be further complicated by crowded scenes in the real world, e.g. due to occlusions arising in such scenarios. Lim et al. [150] proposed a method using prediction in order to estimate such occlusions. This is followed by the construction of a visibility interval for each capture, which is defined complementary to the occlusion moment, being the input for the subsequent scheduling mechanism using a greedy graph search method.

In contrast to our work, all of the aforementioned approaches suffer from the fact that the image acquisition process cannot be controlled in terms of the camera's (x,y,z)-position. The mobility is restricted to panning or tilting. Additionally, the overall performance in terms of wide-area target acquisition has not been investigated, since sophisticated collaboration and coordination mechanisms are needed for this purpose. Similar to our work, Krahnstoever et al. [135] model the imaging process by utilizing quality functions considering the distance and view angle of targets. Nevertheless, their system is based on PTZ cameras and a centralized architecture. Therefore, their system is not scalable and evaluated in terms of four cameras only. Our reconfiguration method has been evaluated by simulations of up to 100 Active Cameras.

Since all of the approaches mentioned before become non applicable as the number of targets increases and exceeds the number of PTZ cameras, methods based on active vision agents have been proposed, which are presented in the following section. Due to their ability to change their position as well as orientation at runtime, they are able to compensate this drawback by means of coordination.

7.3.3 Active Vision Agents

Active vision agents are a relatively new research field. Since the number of participating cameras increases, the use of a centralized planner for dynamic camera placement and selection becomes more difficult. Due to this reason, a number of agent-based approaches has emerged in the literature, trying to address the problem of on-line camera planning in order to decrease complexity and increase robustness and scalability. These agents are reconfigurable in terms of their position and orientation. They are based on the idea of *distributed vision networks*, where the participating nodes cooperate to achieve the system's objective. Real-world scenarios are usually motivated by application scenarios of mobile robots or autonomous vehicles.

In the past, multi-agent planning and negotiation techniques have become common in the *Artificial Intelligence* literature but have not been integrated into computer vision work. Some research has focused on developing centralized and distributed methods for multi-agent planning specifically in the context of unmanned vehicles [155, 156]. For example, multiple mobile robots (autonomous vehicles) cooperate with each other to fulfill their navigation tasks in [157], which means that each mobile robot plans its path based on other robots' navigation information.

In [158], the problem of tracking and estimating the motion of a moving target is investigated. For this purpose, a team of mobile robots is used. Each robot is equipped with a directional sensor (e.g. a camera) with limited range. A sensor fusion scheme based on the interrobot communication is proposed to obtain the accurate real-time information of the target's position and motion. Betser et. al. [159] present a method for controlling several unmanned autonomous vehicles flying in formation and utilizing visual information. Flying vehicles in following mode track the leaders via visual means. In [160], Ng et. al. present a method, which is able to coordinate the movements of multiple robots in order to follow a search tactic in a collective manner. This is performed in an unknown and cluttered environment. First, individual robot reactive behaviors are developed, which allow for coordinated movements. Each robot is programmed with the same set of primitive behaviors: (1) obstacles negotiation, (2) homing, (3) flocking, and (4) searching, with obstacles negotiation being the most important and searching being the least important step. According to different environmental stimulants, the robots adopt one of these behaviors at a time according to their order of importance for the cooperation purposes.

In summary, approaches in this research domain focused on planning for navigation of the vehicles and not on investigating mechanisms for dynamic reconfiguration to support underlying computer vision algorithms, i.e. the wide-area target acquisition problem, which we considered in this thesis.

7.4 Time Synchronization in Sensor Networks

In the last years, various time synchronization schemes have been developed for the use in wireless sensor networks [67]. One feature that can be used to classify them is to examine whether a time synchronization scheme uses a *sender-to-receiver* or a *receiver-to-receiver* approach. The former, rather traditional approach, is used by traditional time synchronization schemes such as NTP [50]. Receiver-to-receiver approaches have recently gained more interest by the research community, since they support system scalability [67].

7.4. TIME SYNCHRONIZATION IN SENSOR NETWORKS

Traditional synchronization algorithms are based on the estimation of transmission times. Any transmission of a network packet can be divided into four phases with each taking a different amount of time, as it is described by v. Greunen and Rabaey [161]:

Send time The time spent on assembling the message at the sender, which includes processing and buffering time.

Access time The delay associated with accessing the channel.

Propagation time The time for the signal to propagate across the physical medium between the two nodes.

Receive time The processing time required to receive the message from the channel and notify the host of its arrival.

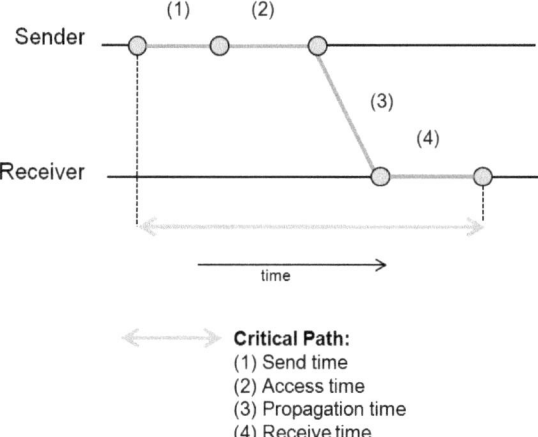

Figure 7.2: Critical path of traditional time synchronization protocols

Together, these four phases make up the so-called *time-critical path*, which is the path of a message that contributes to non-deterministic errors in the protocol like message delays [67]. On the one hand, the uncertainty of the send time, access time and receive time is driven by the uncertainty of the hardware and hardware settings of the network nodes which can be a major problem in heterogeneous networks or pervasive applications. On the other hand, the propagation time depends significantly on the communication medium that might depend on the scenario and external influences.

Most existing methods synchronize a sender with a receiver by transmitting the current clock values as time stamps. As a consequence, these methods are vulnerable to variances in message delay. Using the traditional sender-to-receiver approach involves the estimation of the whole transmission time of sender packets as depicted in Figure 7.2. Newer methods perform synchronization among receivers using the time at which each of them receives the same message. Such an approach reduces the time-critical path, which is the path of a message that contributes to non-deterministic errors in the protocol. Nevertheless, the sender-to-receiver synchronization method is claimed to be more precise than the receiver-to-receiver synchronization.

The following two subsections present the state of the art in both domains.

7.4.1 Sender-to-Receiver Synchronization

Principally, sender-to-receiver approaches involve some sender transmitting network packets to receiver nodes. For instance, these packets contain a reference time or are used in some other way for synchronization purposes.

Genlock and NTP Genlock [162] is a technique that is preferably used for frame synchronization of video sources like cameras. NTP [50] is preferably used in computer networks. Both approaches are based on a centralized approach. With the help of Genlock, it is possible to synchronize television picture sources like cameras, video recorders or digitizers by using an external synchronization signal. All connected television picture sources use that so-called *genlock signal* as a kind of central clock and adapt their vertical, horizontal, frame and color synchronization to create a proper standard composite video signal. NTP enables clock synchronization of computer systems

7.4. TIME SYNCHRONIZATION IN SENSOR NETWORKS 177

by periodically transmitting network packets with timing information. The underlying structure of NTP is hierarchical and is composed of different strata of network nodes. At the top of this hierarchical structure, atomic and radio-controlled clocks are located which provide the correct reference time. The network nodes in the bottom strata represent the end-users or, in our case, the Active Cameras.

Sensor Networks Several researchers proposed message-based sender-to-receiver approaches for sensor networks in the recent years [67].

The protocol proposed by Römer et al. has been implemented for the use in mobile ad-hoc networks and is based on an innovative time transformation algorithm for achieving clock synchronization [163]. This protocol is especially effective in environments with strict resource constraints. Nodes can either be in the role of being a sender or a receiver. A sender transmits a message and attaches a specific time stamp to each message. This time stamp contains a time interval. Synchronization is performed pairwise between nodes (a sender and a receiver) whenever time stamps are exchanged.

Ping et al. [164] developed a protocol for maintaining a uniform notion of time among nodes that participate in a network. A global time stamp provides the basis for merging individual sensor readings into a database. For this purpose, a leader is chosen among a set of communicating nodes and this leader broadcasts its local clock value to the other nodes. All receiving nodes compare their local clock values relative to the leader's time and if the delays in the path from one node to another node can be estimated accurately, the two nodes can be synchronized. When the message is broadcasted, the sender of the packet will be synchronized with all nodes receiving its packet.

Li and Rus have defined a so-called *rate-based diffusion protocol* in which nodes achieve synchronization by flooding their neighbors with information about each node's local clock value [165]. After each node has learned the clock values of all its neighbors, the node can use a mutually agreed consensus value to adjust its clock. Examples of consensus values suggested by the authors include the highest clock reading in the net, the lowest clock reading, or some statistical value based on the clock readings (e.g. the average or the median of the readings). According to the authors, using the highest or the lowest reading yields the simplest synchronization algorithm; however, this strategy lacks robustness. A malicious or erratic node may impose an abnormally high (or low) clock value on the whole network.

In contrast to our method, all aforementioned sender-to-receiver approaches use message-based synchronization schemes and timing information of these messages. Usually, they are based on the assumption of symmetric round-trip times. Our approach does not depend on these timing information, since it utilizes visual events. Therefore, timing information of the underlying communication network are irrelevant for our approach and we are not prone to disturbances (e.g. network latencies) influencing this timing information.

7.4.2 Receiver-to-Receiver Synchronization

The receiver-to-receiver synchronization exploits the property of the physical broadcast medium that if any two nodes receive the same message in single-hop transmission, the message arrives at approximately the same time. Instead of interacting with a sender, nodes exchange the time at which they received the same message and compute their offset based on the difference in reception times. The obvious advantage is the reduction of the message-delay variance. These protocols are only vulnerable to the propagation delay to the various receivers and the differences in receive time.

Video Synchronization and Visual Temporal Calibration In [166], Polleyfeys et al. use space-time interest point distributions for video synchronization. They correlate space-time interest points between videos and show that by detecting, selecting, and correlating the distribution of space-time interest points, videos from different viewpoints can be automatically synchronized. An approach in the Fourier domain has been presented by Kuthirummal et al. [167], which needs to compute weak calibration in the form of the trilinear tensor before alignment, thus requiring at least seven stationary corresponding points in three views. Additionally, a point needs to be tracked over a number of frames in three views. Lee et al. [168] use

7.4. TIME SYNCHRONIZATION IN SENSOR NETWORKS

geometric constraints to align the tracking data in time. This method requires knowledge about the intrinsic camera parameters. Accuracy can be affected by the height of the objects and the object-to-camera distance. Velipasalar et al. [169] describe a method for temporally calibrating video sequences from unsynchronized cameras by image processing operations. They propose a method in which foreground objects are tracked. A point of interest is extracted for each object as its current location, and the corresponding location of the object in the other sequence is obtained by using projective invariants

All of the aforementioned algorithms are based on centralized system architectures, which cannot be guaranteed in Active Camera Networks. Additionally, their methods do not contain active optical elements (such as an optical beacon) for supporting synchronization based on visual cues.

Sensor Networks One of the most popular representatives in the receiver-to-receiver domain is the reference broadcast synchronization algorithm (RBS) [170]. In case of RBS, a transmitter (e.g. a time server) broadcasts a reference packet to multiple receivers. Each receiver records the time at which the reference packet is received according to its local clock. Afterwards, these observed times are exchanged between the receivers. The clock offset between two receivers is computed as the difference between the local times at which the nodes received the same message [67]. Using RBS eliminates the uncertainty of the sender by its removal (including the send and access time) from the critical path.

The probabilistic clock synchronization service in sensor networks defined by PalChaudhuri et al. [171] extends RBS by providing means to set protocol parameters. Thus, it is possible to derive an appropriate number of messages used for clock offset and clock skew estimation between the local clocks from probabilistic models. This gives a better control on the synchronization overhead. Mock et. al. [172] defined a protocol for continuous clock synchronization in wireless sensor networks by extending the IEEE 802.11 standard for wireless local area networks. Though the design of their protocol is different from RBS, Mock et al. also exploit the tightness of the communication medium when using reference broadcasts. Quite different is the time-diffusion synchronization protocol (TDP) [173]. TDP contains different algorithms and induces a common notion of system-wide time with the help of cyclically executed diffusions of timing messages that help the several nodes to converge their local times. The self-organizing features of this protocol are very distinctive and even allow the establishment of a common notion of time without external time servers. Nevertheless, it is also prone to errors due to link-asymmetry in terms of the communication link.

All receiver-to-receiver synchronization protocols above have in common that they are based on the assumption of the tightness of the communication medium. Nevertheless, this assumption does not hold in all cases for network communication. RBS, for example, assumes that the reference packet is received by multiple receivers nearly simultaneously and the propagation time of each received message is the same. Nevertheless, this assumption might not be tenable in application scenarios where the communication medium is exposed to severe external influences and hence the propagation time is very uncertain. For instance, this is the case in forest fire surveillance scenarios [174]. Finally, the propagation time is a function of the distance and can vary strongly due to motion of network nodes. This drawback is overcome by our method of active frame synchronization by utilizing visual events for time synchronization.

7.5 Summary

The aforementioned research fields cover aspects of dynamic reconfiguration. Figure 7.3 shows an overview of selected related works and the abilities they lack in comparison to the methods presented in this thesis.

Real-time systems utilize dynamic reconfiguration at the scheduler level. Thus, they make way for flexibility and adaptivity. Nevertheless, tasks are often assumed to have the same periods and deadlines, which is not an acceptable assumption for dynamic environments. In addition, the reconfiguration methods usually run in a single scheduler. In contrast to our methods, they lack from scalability and applicability for decentralized system structures.

The Dynamic Vehicle Routing Problem with Time Windows addresses the

7.5. SUMMARY

problem of dynamic reconfiguration, too. Vehicles have to be routed and scheduled to customers without exceeding temporal and spatial constraints. In contrast to our work, these approaches usually assume the on-site service time to be zero. The on-site service is assumed to be non-zero for applicability in our work, e.g. for computer vision algorithms. In addition, we assume the expiration time of the targets (i.e. customers in the vehicle routing problem domain) to be dynamic, since it depends on the velocity of the targets (e.g. humans, moving through the surveillance area).

In the realm of sensor planning for visual surveillance, dynamic reconfiguration is utilized for active sensor planning to determine the configuration of a set of sensors in static or dynamic surveillance environments. In contrast to our methods, sensing performances of proposed systems do usually not take the requirements of the underlying computer vision algorithm into account.

Furthermore, we reviewed the state of the art in the field of time synchronization. One feature that can be used to classify approaches in this domain is to examine whether a time synchronization scheme uses a *sender-to-receiver* or a *receiver-to-receiver* approach. The former, rather traditional approach, is used by traditional time synchronization schemes such as NTP. Receiver-to-receiver approaches have recently gained more interest by the research community, since they support system scalability.

In contrast to our method, the reviewed sender-to-receiver approaches use message-based synchronization schemes and timing information of these messages. They are based on the assumption of symmetric round-trip times. Our approach does not depend on these timing information, since it utilizes visual events. Therefore, our approach is not prone to disturbances (e.g. network latencies) influencing the symmetry of round-trip times. The presented receiver-to-receiver synchronization protocols are based on the assumption of the tightness of the communication medium. This assumption might not be tenable in applications where the communication medium is exposed to severe external influences and hence the propagation time is very uncertain, as given in Active Camera Networks. Finally, the propagation time is a function of the distance and can vary strongly due to motion of network nodes. This drawback is overcome by our method by utilizing visual events for time synchronization.

	Dynamic Reconfiguration						Time Synchronization		
	Scalable	Support of aperiodic Tasks	Graceful Degradation	Consideratin of on-site Service	Support of Computer Vision	Position Management	Sender-to-Receiver	Receiver-to-Receiver	Use of Visual Events
Jehuda et al.	no	no	yes	no	no	no			
Rusu et al.	no	no	no	no	no	no			
Lima et al.	no	no	no	no	no	no			
Pavone et al.	no	yes	yes	no	no	yes			
Matsuyama et al.	Up to 10 nodes	yes	yes	no	no	yes			
Horling et al.	Up to 10 nodes	yes	yes	no	no	yes			
Bakhtari et al.	Up to 10 nodes	yes	yes	no	no	yes			
Hörster et al.	no	no	no	no	no	no			
Krahnstoever et al.	no	yes	yes	yes	yes	no			
Ng. et al.	Up to 10 nodes	yes	yes	no	no	yes			
Mills et al.							yes	no	no
Römer et al.							yes	no	no
Elson et al.							no	yes	no
Polleyfeys et al.							no	yes	yes
Velipasalar et al.							no	yes	yes
Wittke	At least up to 100 nodes	yes	yes	yes	yes	yes	yes	yes	yes

Figure 7.3: Overview of related work

Chapter 8

Conclusion

Based on advances in the research areas of robotics and computer vision, this work introduces a system architecture that serves as a basis for Active Camera Networks. Applications of such systems are manifold and include, for example, the exploration and surveillance of large areas. Key components to robustly implement such applications are dynamic reconfiguration methods for distributed control that adapt the system's behavior to changing environmental conditions and efficiently coordinate the usage of system resources, in particular the available cameras. Currently, passive camera networks try to cover the entire area or at least the most important parts of it. Therefore, system designers have to determine the number of cameras and their placements based on *a priori* information considering the requirements of the underlying surveillance task, e.g. number and frequency of targets occurring and so-called *hot spots* of occurrence. This approach is applicable to controlled and static environments. Nevertheless, *a priori* information becomes less useful in dynamic environments, since dynamics such as a varying number of targets may occur at runtime. The broad range of requirements, which algorithms for the interpretation of scenes from multiple perspectives have, adds up to these difficulties and again increases the number of cameras that are necessary. One viable and cost-effective alternative to just increasing the number of cameras to the demands of surveillance applications is to make efficient use of Active Camera Networks and dynamic reconfiguration methods as presented in this thesis. These methods enable Active Cameras to collaborate in surveillance scenarios by relying on Organic Computing principles such as self-configuration, self-organization, and self-adaptation.

8.1 Summary of Contributions

Two reconfiguration methods have been investigated and analyzed thoroughly. They allow for self-organizing wide-area target acquisition and self-configuring frame synchronization based on visual events. Special prerequisites have been considered to enable a coping of the methods with dynamic environments such as moving observation targets. An evaluation with a realistic model in terms of computer vision and camera repositioning shows that these methods are suitable for Active Camera Networks consisting of up to 100 Active Cameras. They are robust towards common disturbances in dynamic environments and exceed static camera systems in terms of flexibility and efficiency.

Various conclusions can be drawn in the end of this work. In order to build Active Camera Networks that are equipped with organic system properties such as self-organization, self-adaptation, and self-configuration, several issues have to be considered. As stated at the beginning of this thesis, three main problems have been addressed in this thesis:

- **System architecture:** Active Camera Networks, as introduced in Section 2, comprise a high number of Active Cameras in order to cooperatively solve surveillance tasks, which could not be achieved by a single camera or only through considerably more stationary cameras. For this purpose, the camera's mobility plays an essential role and has to be managed regarding the camera's orientation and position. A concept for implementing realistic sensing performance metrics has been introduced and integrated into the dynamic reconfiguration methods. Thus, the requirements of the underlying computer vision algorithms have been integrated into the sensing constraints. Furthermore, a distributed software

8.1. SUMMARY OF CONTRIBUTIONS

architecture has been presented, which can be executed on each camera independently. This distributed design allows for scalability regarding the number of cameras and adaptability in terms of the environment.

- **Wide-area target acquisition:** Based on a suitable system architecture, the problem of wide-area target acquisition has been discussed. Solutions to this problem, which is NP-complete (see Section 5.1.2), can be approximated by the distributed control heuristic *DRofACN*. This method has been developed to allow for dynamic and distributed control of nodes in Active Camera Networks. It is based on the objective to capture high quality images of moving targets. Application scenarios have been investigated where events unfold over a large geographic area and close-up views are acquired for biometric tasks such as face detection. Utilizing Active Cameras in such a scenario makes way for efficient use of resources. Nevertheless, this control cannot be based on separate analysis of the sensed imagery in each camera. They act collaboratively to be able to acquire exactly one capture of each target. Simulations with up to 100 Active Cameras show the scalability and reliability of the proposed method. The performance of different target generation rates is analyzed and it is shown that an Active Camera Network of 100 nodes can handle up to 2,500 targets simultaneously with a detection rate of 90 % and a mean target detection time of less than 10 seconds.

- **Active frame synchronization:** In order to pave the way for self-configuration in Active Camera Networks, novel algorithms for frame-level and visual cue-based clock synchronization have been developed called *ACFSync*. Thus, no centralized time synchronization server is needed any more. Two methods have been proposed in Section 5.3: (1) a clock synchronization method based on an optical beacon, and (2) a synchronization method where Active Cameras sharing the same field of view can synchronize their clocks cooperatively. One aspect that both methods have in common is that they utilize optical events. Hence, they do not rely on specific hardware. The beacon-based approach achieves a synchronization accuracy of one frame in 70 % of the cases. If participating cameras capture their environment with a frame rate of 25 frames per second, the active approach achieves an accuracy of 40 ms. A synchronization accuracy within milliseconds is sufficient for scenarios where visual events are triggered by human beings moving with a velocity in the order of few meters per second. It has been shown that the second approach achieves good synchronization results if the difference of the view angles of both cameras is less than 45° or if they are counterpart.

With the reconfiguration methods presented in this thesis, a toolkit for building Active Camera Networks has been introduced. All methods have been evaluated in scenarios derived from real-world applications. For the algorithm *DRofACN*, presented in Section 5.2, a main station's front yard has been considered. This wide-area set-up has to be protected against potentially dangerous situations induced by humans. Simulation experiments show that collaborating Active Cameras are well suited for this task. The target acquisition ratio becomes close to optimal so that an image exceeding a pre-defined minimum quality is captured for the majority of relevant targets. Based on the number of Active Cameras used, the average target detection time can be reduced to less than 10 seconds.

The cooperative frame synchronization method has been simulated with realistic 3-D models of persons moving through a camera's field of view. Multiple cameras capture this scene and synchronize their camera clocks on the basis of visual cues derived from this video sequence. The impact of the number of persons within the video sequence as well as their perspective's offset have been investigated. On the basis of an optical beacon that has been developed, Active Cameras can synchronize their clock on their own. For this purpose, they have to decode an optical message containing a time stamp. The efficiency and correctness of this method has been investigated in real-world experiments. With this thesis, dynamic reconfiguration methods have been presented allowing for self-organization and self-optimization in Active Camera Networks even in large systems with up to 100 cameras.

8.2 Future Research Opportunities

Future research opportunities arise based on the reconfiguration methods presented in the previous chapters. For both reconfiguration methods presented in this thesis, ideas for future research are discussed in the following.

Wide-Area Target Acquisition Future work might focus on extending *DRofACN* by more intelligent, i.e. self-learning, mechanisms for phenomenon adaptability, since classical surveillance scenarios are usually characterized by high- and low-activity zones. Adding dynamic and adaptive actuation ranges to *DRofACN* can balance this load. For this purpose, dominant motions have to be learned from the past and cameras have to self-adapt their actuation ranges in order to build spatial redundancy zones to balance the load of target observation between neighboring cameras. In addition, the size of the actuation ranges could be increased. Nevertheless, network partitions could arise in this case, which have to be handled by routing mechanisms. Furthermore, the current mechanism for phenomenon adaptability (*ENRA*, see Section 5.2.6) could be investigated in terms of optimal configuration parameters, e.g. how the adaptability of the algorithm is influenced by varying the time to update parameter. As presented in Section 5.2, *DRofACN* makes use of imaging quality functions to achieve optimal imaging conditions. Optimality is achieved by maximizing the probability of successfully completing the addressed vision task (e.g. a biometric task), which is determined by an objective function according to capturing conditions. These conditions contain the distance at which the target is imaged and the view angle. Nevertheless, more parameters could be modeled on the basis of this function. On example is given by the estimated deadlines of targets by which they leave the surveillance area. Additionally, parameters could be added for maximizing the observation time of targets across multiple cameras for creating multi-view scenarios for object reconstruction. The function might also consider obstacles, lighting conditions or other workspace-specific information.

Beacon-based Clock Synchronization A reconfiguration method for active frame synchronization based on an optical beacon has been introduced in Section 5.3.3. This method can be enhanced to achieve an increased synchronization performance and accuracy. The specific shape of the waveform used to modulate the brightness signal is a rectangle wave which does not need to be the optimum. The implemented line encoding is based on codes used in hard-drives during the 1970ies. Therefore, more current encoding schemes could potentially be used here, for example, Turbo Codes or low-density parity-check (LDPC) codes. The beacon signal could also be used as smart landmarks for self-localization of autonomous mobile robots. Thus, traditional landmarks for self-localization could be enriched with additional information. Additional colors or timing information could be investigated to increase the information density of the beacon signal.

Cooperative Frame Synchronization A reconfiguration method for cooperative frame synchronization has been presented in Section 5.3.4. On the basis of the optical flow motion field, saliency curves are computed and correlated. Other low-level features such as brightness values of the picture could be integrated to increase the robustness of the method. In addition to low-level features, high-level features such as detected objects within the image could also serve as input. Since the saliency measure is independent from the data types, heterogeneous sensor data could be utilized. One example is given by data stemming from audio or temperature sensors or other optical sensors such as thermal cameras. Furthermore, IP camera networks suffer from the problem of dynamics concerning the image capturing process. Thus, synchronized cameras can acquire image sequences asynchronously, e.g. due to maximum CPU usage or changing lighting conditions. This asynchrony might be detected and corrected by a cooperative frame synchronization method.

This list of research opportunities is way too short to cover all possible issues that need to be investigated in future. It contains some ideas that arose in the context of this thesis and might be considered in upcoming research projects. Especially, a lot of work still needs to be done in the fields of collaborative behavior for multi-vehicle systems and active vision.

Bibliography

[1] Michael D. Naish, Elizabeth A. Croft, and Beno Benhabib, "Coordinated dispatching of proximity sensors for the surveillance of maneuvering targets," in *Proceedings of Robotics and Computer Integrated Manufacturing*, 2003, vol. 3, pp. 283–299. 5, 27, 164

[2] Fred W. Rauskolb, Kai Berger, Christian Lipski, Marcus Magnor, Karsten Cornelsen, Jan Effertz, Thomas Form, Fabian Graefe, Sebastian Ohl, Walter Schumacher, Jörn-Marten Wille, Peter Hecker, Tobias Nothdurft, Michael Doering, Kai Homeier, Johannes Morgenroth, Lars Wolf, Christian Basarke, Christian Berger, Tim Gülke, Felix Klose, and Bernhard Rumpe, "Caroline: An autonomously driving vehicle for urban environments," *Journal of Field Robotics*, vol. 25, pp. 674–724, September 2008. 5, 30, 31

[3] Bernhard Rinner and Markus Quaritsch, "Embedded middleware for smart camera networks and sensor fusion," in *Multi-Camera Networks - Principles and Applications*, Academic Press, Ed. 2010, pp. 511–537, Elsevier. 5, 34, 35, 56

[4] Ardevan Bakhtari, "Multi-target surveillance in dynamic environments: Sensing-system reconfiguration," *Ph.D. thesis, Department of Mechanical and Industrial Engineering, University of Toronto*, 2006. 5, 39

[5] Leila De Floriani and Paola Magillo, "Algorithms for visibility computation on terrains: a survey," *Environment and Planning B: Planning and Design*, vol. 30, no. 5, pp. 709–728, 2003. 5, 47, 48

[6] DoHyung Kim, Woo han Yun, and Jaeyeon Lee, "Tiny frontal face detection for robots," in *Proceedings of 3rd International Conference on Human-Centric Computing (HumanCom)*, August 2010, pp. 1–4. 6, 63, 81

[7] Sooraj Kumar and Andreas Savakis, "Face recognition with variation in pose angle using face graphs," M.S. thesis, Department of Computer Engineering, Kate Gleason College of Engineering, Rochester Institute of Technology, Rochester, NY, 2009. 6, 82

[8] Fritz Webering, "Beacon-assisted optical clock synchronization in smart camera networks," B.S. thesis, Institute of Systems Engineering / SRA, Leibniz Universität Hannover, September 2010. 8, 103, 136

[9] Douglas C. Schmidt, "Middleware for real-time and embedded systems," *Communications of the ACM – Adaptive middleware*, vol. 45, pp. 43–48, June 2002. 9, 166

[10] Hartmut Schmeck, "Organic Computing - Vision and Challenge for System Design," in *PARELEC*, 2004, p. 3. 13

[11] Stephan Hengstler and Hamid Aghajan, "A Smart Camera Mote Architecture for Distributed Intelligent Surveillance," in *Working Notes of the International Workshop on Distributed Smart Cameras (DSC)*, 2006. 13

[12] Michael Wittke and Jörg Hähner, "Self-organising distributed smart camera systems," in *Organic Computing - A paradigm shift for complex systems*, Christian Müller-Schloer, Hartmut Schmeck, and Theo Ungerer, Eds., pp. 609–610. Birkhäuser Verlag, 2011. 13

[13] Web, "MUViT: Mustererkennung und Video Tracking: Sozialpsychologische, soziologische, ethische und rechtswissenschaftliche Analysen," *http://www.bmbf.de/de/14395.php*, 2010. 16

[14] Sergio A. Velastin, Boghos A. Boghossian, Benny P. L. Lo, Jie Sun, and Maria A. Vicencio-Silva, "PRISMATICA: toward ambient intelligence in public transport environments," *IEEE Transactions on Systems, Man*

and *Cybernetics, Part A: Systems and Humans*, vol. 35, no. 1, pp. 164–182, January 2005. 16

[15] Jürgen Branke, Moez Mnif, Christian Müller-Schloer, and Holger Prothmann, "Organic Computing - Addressing Complexity by Controlled Self-Organization," in *Proceedings of Second International Symposium on Leveraging Applications of Formal Methods, Verification and Validation (ISoLA)*, November 2006, pp. 185–191. 17

[16] Michael J. Swain and Markus A. Stricker, "Promising directions in active vision," *International Journal of Computer Vision*, vol. 11, pp. 109–126, 1993. 24, 25, 29

[17] Michael Bramberger, Andreas Dobl, Arnold Maier, and Bernhard Rinner, "Distributed embedded smart cameras for surveillance applications," *Computer*, vol. 39, pp. 68–75, 2006. 24

[18] John G. Webster, *The measurement, instrumentation and sensors handbook*, CRC Press, 1999. 28

[19] D. M. Gavrila and S. Munder, "Multi-cue pedestrian detection and tracking from a moving vehicle," *International Journal of Computer Vision*, vol. 73, pp. 41–59, June 2007. 28

[20] Ambareen Siraj, Rayford B. Vaughn, and Susan M. Bridges, "Intrusion sensor data fusion in an intelligent intrusion detection system architecture," in *Proceedings of Hawaii International Conference on System Sciences*, vol. 9, 2004. 28

[21] Eduardo Monari, Jochen Maerker, and Kristian Kroschel, "A robust and efficient approach for human tracking in multi-camera systems," in *Proceedings of the Sixth IEEE International Conference on Advanced Video and Signal Based Surveillance (AVSS)*, Washington, DC, USA, 2009, pp. 134–139, IEEE Computer Society. 28

[22] Mark Brown, "Technology: Google has travelled 140,000 miles in self-driving cars," http://www.wired.co.uk/news/archive/2010-10/11/google-self-driving-cars, October 2010. 31

[23] Les Dorr and Alison Duquette, "Fact sheet - unmanned aircraft systems (uas)," http://www.faa.gov/news/fact_sheets/, December 2010. 32

[24] Markus Quaritsch Benrhard Rinner, "Toward pervasive smart camera networks," in *Multi-Camera Networks - Principles and Applications*, Academic Press, Ed. 2010, pp. 483–496, Elsevier. 34

[25] C. B. Margi, X. Lu, G. Zhang, G. Stanek, and K. Obraczka, "Meerkats: A power-aware, self-managing wireless camera network for wide area monitoring," in *Proceedings of the International Workshop on Distributed Smart Cameras*, 2006. 36, 37

[26] Mohammad Rahimi, Rick Baer, Obimdinachi I. Iroezi, Juan C. Garcia, Jay Warrior, Deborah Estrin, and Mani Srivastava, "Cyclops: in situ image sensing and interpretation in wireless sensor networks," in *Proceedings of 3rd ACM Conference on Embedded Networked Sensor Systems (SenSys)*, November 2005. 36, 37

[27] Stephan Hengstler, Daniel Prashanth, Sufen Fong, and Hamid Aghajan, "Mesheye: a hybrid-resolution smart camera mote for applications in distributed intelligent surveillance," in *Proceedings of the 6th International Conference on Information Processing in Sensor Networks (IPSN)*, New York, NY, USA, 2007, pp. 360–369, ACM. 36, 37

[28] Richard Kleihorst, Anteneh Abbo, Ben Schueler, and Alexander Danilin, "Camera mote with a high-performance parallel processor for real-time frame-based video processing," in *Proceedings of the IEEE Conference on Advanced Video and Signal Based Surveillance (AVSS)*, Washington, DC, USA, 2007, pp. 69–74, IEEE Computer Society. 36, 38

[29] Anthony Rowe, Adam G. Goode, Dhiraj Goel, and Illah Nourbakhsh, "CMUcam3: An open programmable embedded vision sensor," Tech.

Rep. CMU-RI-TR-07-13, Robotics Institute, Pittsburgh, PA, May 2007. 36, 38

[30] Phoebus Chen, Parvez Ahammad, Colby Boyer, Shih-I Huang, Leon Lin, Edgar Lobaton, M. Lenore Meingast, Songhwai Oh, Simon Wang, Posu Yan, Allen Yang, Chuohao Yeo, Lung-Chung Chang, Doug Tygar, and S. Shankar Sastry, "Citric: A low-bandwidth wireless camera network platform," in *Proceedings of the Second ACM/IEEE International Conference on Distributed Smart Cameras (ICDSC)*, September 2008, pp. 1–10. 36, 38

[31] Gary Bradski, "The OpenCV Library," *Dr. Dobb's Journal of Software Tools*, 2000. 39, 47, 60

[32] Roger Y. Tsai, "A versatile camera calibration technique for high-accuracy 3d machine vision metrology using off-the-shelf tv cameras and lenses," *IEEE Journal of Robotics and Automation*, pp. 221–244, 1992. 40

[33] Zhaolin Cheng, Dhanya Devarajan, and Richard J. Radke, "Determining vision graphs for distributed camera networks using feature digests," *EURASIP Journal on Advances in Signal Processing*, vol. 2007, no. 1, pp. 220–231, 2007. 40, 48

[34] Kentaro Toyama, John Krumm, Barry Brumitt, and Brian Meyers, "Wallflower: Principles and practice of background maintenance," in *Proceedings of the Seventh IEEE International Conference on Computer Vision*, Los Alamitos, CA, USA, 1999, vol. 1, pp. 255–261, IEEE Computer Society. 40

[35] Berthold K. P. Horn and Brian G. Schunck, "Determining optical flow," *ARTIFICAL INTELLIGENCE*, vol. 17, pp. 185–203, 1981. 41, 105

[36] Erdogan Dur, "Optical flow-based obstacle detection and avoidance behaviors for mobile robots used in unmanned planetary exploration," in *Proceedings of the 4th International Conference on Recent Advances in Space Technologies (RAST)*, June 2009, pp. 638–647. 41

[37] Paul Viola and Michael Jones, "Robust real-time face detection," *International Journal of Computer Vision*, vol. 57, pp. 137–154, 2004. 41

[38] Q. Cai and J.K. Aggarwal, "Tracking human motion in structured environments using a distributed-camera system," *IEEE Transactions on Pattern Analysis and Machine Intelligence*, vol. 21, no. 11, pp. 1241–1247, November 1999. 42

[39] Ting-Hsun Chang and Shaogang Gong, "Tracking multiple people with a multi-camera system," in *IEEE Workshop on Multi Object Tracking*, 2001, pp. 19–26. 42

[40] Robert Collins, Alan Lipton, and Takeo Kanade, "A system for video surveillance and monitoring," in *American Nuclear Society 8th Internal Topical Meeting on Robotics and Remote Systems*, 1999. 42

[41] Timothy Huang and Stuart Russell, "Object identification in a bayesian context," in *Proceedings of the Fifteenth International Joint Conference on Artificial Intelligence*. 1997, pp. 1276–1283, Morgan Kaufmann. 42

[42] Robert T. Collins, Omead Amidi, and Takeo Kanade, "An active camera system for acquiring multi-view video," in *Proceedings of the International Conference on Image Processing*, 2002, pp. 517–520. 42

[43] Uwe Jänen, Christian Paul, Michael Wittke, and Jörg Hähner, "Multi-object tracking using feed-forward neural networks," in *Proceedings of the International Conference on Soft Computing and Pattern Recognition (SoCPaR)*, Paris, France, 2010, pp. 176–181, IEEE Computer Society. 42

[44] Andreas Koschan, "What is new in computational stereo since 1989: A survey on current stereo papers," Tech. Rep., Technische Universität Berlin and Technischer Bericht, 1993. 42

[45] Enrico Grosso and Massimo Tistarelli, "Active/dynamic stereo vision," *IEEE Transactions on Pattern Analysis and Machine Intelligence*, vol. 17, pp. 1117–1128, November 1995. 42

[46] Antii Kotanen, Marko Hannikainen, Helena Leppakoski, and D. Timo Hamalainen, "Positioning with IEEE 802.11b wireless LAN," in *Proceedings of IEEE Personal, Indoor and Mobile Radio Communications (PIMRC)*, September 2003, vol. 3, pp. 2218–2222. 46

[47] Alan T. Murray, Kamyoung Kim, James W. Davis, Raghu Machiraju, and Richard E. Parent, "Coverage optimization to support security monitoring," *Computers, Environment and Urban Systems*, vol. 31, no. 2, pp. 133–147, 2007. 48

[48] Uğur Murat Erdem and Stan Sclaroff, "Automated camera layout to satisfy task-specific and floor plan-specific coverage requirements," *Computer Vision and Image Understanding - Special issue on omnidirectional vision and camera networks*, vol. 103, no. 3, pp. 156–169, 2006. 48

[49] Kay Römer, Philipp Blum, and Lennart Meier, "Time synchronization and calibration in wireless sensor networks," in *Handbook of Sensor Networks: Algorithms and Architectures*, Ivan Stojmenovic, Ed., pp. 199–237. John Wiley & Sons, September 2005. 50

[50] David L. Mills, *Internet Time Synchronization: The Network Time Protocol*, RFC Editor, United States, 1989. 51, 69, 174, 176

[51] Richard E. Schantz and Douglas C. Schmidt, "Research Advances in Middleware for Distributed Systems," in *Proceedings of the IFIP 17th World Computer Congress - TC6 Stream on Communication Systems: The State of the Art*, Deventer, The Netherlands, 2002, pp. 1–36, Kluwer, B.V. 57

[52] Michael Wittke, Carsten Grenz, and Jörg Hähner, "Towards organic active vision systems for visual surveillance," in *Architecture of Computing Systems - ARCS 2011*, Mladen Berekovic, William Fornaciari, Uwe Brinkschulte, and Cristina Silvano, Eds., vol. 6566 of *Lecture Notes in Computer Science*, pp. 195–206. Springer Berlin / Heidelberg, 2011. 57

[53] Monika Sester and H. Neidhart, "Reconstruction of building ground plans from laser scanner data," in *Proceedings of the AGILE, Girona, Spain*, 2008. 59

[54] Aman Kansal, William Kaiser, Gregory Pottie, Mani Srivastava, and Sukhat Gaurav, "Reconfiguration methods for mobile sensor networks," *ACM Transactions on Sensor Networks*, vol. 3, October 2007. 59, 63

[55] Radoslaw Rudnicki, Monika Sester, and Volker Paelke, "Visual interactive exploration of spatio-temporal patterns," in *International Workshop on Visual Languages and Computing, Redwood City, USA*, 2009. 59, 88

[56] Michael Wittke and Jörg Hähner, "Distributed vision graph update in mobile vision networks," in *Workshop Proceedings of the 23th International Conference on Architecture of Computing Systems (ARCS)*, 2010. 59, 67

[57] Michael Wittke, Uwe Jänen, Aret Duraslan, Emre Cakar, Monika Steinberg, and Jürgen Brehm, "Activity recognition using optical sensors on mobile phones," in *Proceedings of GI Jahrestagung*, 2009, pp. 2181–2194. 59, 67

[58] Martin Hoffmann and Jörg Hähner, "ROCAS: A Robust Online Algorithm for Spatial Partitioning in Distributed Smart Camera Systems," in *Proceedings of the First ACM/IEEE International Conference on Distributed Smart Cameras (ICDSC)*, September 2007, pp. 267–274. 62

[59] Lars Friedrichs, Jörg Hähner, Michael Wittke, and Martin Hoffmann, "Method for distributed generation of data for e.g. mobile telephone, in distributed vision network, involves using correlated data for data encryption and/or error detection by detecting devices communicating with each other," *Patent (DPMA)*, , no. 102009005978, 2009. 68

[60] Michael Wittke, Sascha Radike, Carsten Grenz, and Jörg Hähner, "DRofACN: Dynamic Reconfiguration of Active Camera Networks," *Elsevier*

Journal for Computer Communications - Special Issue on Wireless Sensor and Robot Networks: Algorithms and Experiments (submitted), 2011. 69

[61] Richard M. Karp, "Reducibility Among Combinatorial Problems," in *Complexity of Computer Computations*, R. E. Miller and J. W. Thatcher, Eds. 1972, pp. 85–103, Plenum Press. 70

[62] Sergio A. Velastin, Benny P. L. Lo, and Jie Sun, "A flexible communications protocol for a distributed surveillance system," *Journal of Network and Computer Applications*, vol. 27, no. 4, pp. 221–253, 2004. 71

[63] Michail G. Lagoudakis, Evangelos Markakis, David Kempe, Pinar Keskinocak, Anton Kleywegt, Sven Koenig, Craig Tovey, Adam Meyerson, and Sonal Jain, "Auction-based multi-robot routing," in *Robotics: Science and Systems*, 2005, pp. 343–350. 74

[64] Sascha Radike, "Entwurf eines Protokolls zur lokalen räumlichen und zeitlichen Rekonfiguration aktiver Kameras," M.S. thesis, Institute of Systems Engineering / SRA, Leibniz Universität Hannover, April 2011. 76

[65] Mark de Berg, Jörg-Rüdiger Sack, Bettina Speckmann, Anne Driemel, Maike Buchin, Monika Sester, and Marc van Kreveld, "10491 Results of the break-out group: Aggregation," in *Representation, Analysis and Visualization of Moving Objects*, Jörg-Rüdiger Sack, Bettina Speckmann, Emiel Van Loon, and Robert Weibel, Eds., Dagstuhl, Germany, 2011, number 10491 in Dagstuhl Seminar Proceedings, Schloss Dagstuhl - Leibniz-Zentrum fuer Informatik, Germany. 80

[66] Alvaro del Amo Jimenez, "Design of a Protocol for event-based Network Reconfiguration of Active Vision Systems," M.S. thesis, Institute of Systems Engineering / SRA, Leibniz Universität Hannover, April 2011. 87

[67] Bharath Sundararaman, Ugo Buy, and Ajay D. Kshemkalyani, "Clock synchronization for wireless sensor networks: A Survey," *Ad Hoc Networks (Elsevier)*, vol. 3, pp. 281–323, 2005. 90, 174, 176, 177, 179

[68] Claude E. Shannon, "Communication in the presence of noise," *Proceedings of the IRE*, vol. 37, no. 1, pp. 10–21, 1949. 96

[69] John M. Harker, Dwight W. Brede, Robert E. Pattison, George R. Santana, and Lewis G. Taft, "A quarter century of disk file innovation," *IBM Journal of Research and Development*, vol. 25, no. 5, pp. 677–690, 1981. 101

[70] ECMA, "Data Interchange on Read-only 120 mm Optical Data Disks (CD-ROM) second edition," Tech. Rep. ECMA-130, ECMA, June 1996. 101

[71] Shan X. Wang and Aleksandr M. Taratorin, *Magnetic information storage technology*, Academic Press, 1999. 101

[72] Philip Koopman and Tridib Chakravarty, "Cyclic redundancy code (CRC) polynomial selection for embedded networks," in *International Conference on Dependable Systems and Networks*, 2004, pp. 145–154. 103, 104

[73] Bruce D. Lucas, *Generalized Image Matching by the Method of Differences*, Ph.D. thesis, Robotics Institute, Carnegie Mellon University, Pittsburgh, PA, July 1984. 107

[74] Dirk Helbing, Illes Farkas, and Tamas Vicsek, "Simulating Dynamical Features of Escape Panic," *Nature*, vol. 407, pp. 487–490, Sep 2000. 119

[75] Ton Roosendaal and Stefano Selleri, *The Official Blender 2.3 Guide: Free 3D Creation Suite for Modeling, Animation, and Rendering*, No Starch Press, June 2004. 144

[76] Anil M. Cheriyadat and Richard J. Radke, "Detecting dominant motions in dense crowds," *IEEE Journal of Selected Topics in Signal Processing*, vol. 2, no. 4, pp. 568–581, August 2008. 151

[77] Guillem Bernat, Alan Burns, and Albert Llamosi, "Weakly hard real-time systems," *IEEE Transactions on Computers*, vol. 50, pp. 308–321, April 2001. 160

[78] Moncef Hamdaoui and Parameswaran Ramanathan, "A dynamic priority assignment technique for streams with (m, k)-firm deadlines," *IEEE Transactions on Computers*, vol. 44, no. 12, pp. 1443–1451, December 1995. 160

[79] Jair Jehuda and Amos Israeli, "Automated meta-control for adaptable real-time software," *Real-Time Systems*, vol. 14, pp. 107–134, March 1998. 160

[80] Cosmin Rusu, Rami Melhem, and Daniel Mosse, "Multiversion scheduling in rechargeable energy-aware real-time systems," in *Proceedings of the 15th Euromicro Conference on Real-Time Systems*, July 2003, pp. 95–104. 160

[81] George Lima, Eduardo Camponogara, and Ana C. Sokolonski, "Dynamic reconfiguration for adaptive multiversion real-time systems," in *Proceedings of the Euromicro Conference on Real-Time Systems (ECRTS)*, July 2008, pp. 115–124. 161

[82] Karen D. Devine, Erik G. Boman, Robert T. Heaphy, Bruce A. Hendrickson, James D. Teresco, Jamal Faik, Joseph E. Flaherty, and Luis G. Gervasio, "New challenges in dynamic load balancing," *Applied Numerical Mathematics - Adaptive methods for partial differential equations and large-scale computation*, vol. 52, pp. 133–152, February 2005. 161

[83] Harilaos N. Psaraftis, "Dynamic vehicle routing: Status and prospects," *Annals of Operations Research*, vol. 61, pp. 143–164, 1995. 161, 162

[84] Warren B. Powell, Patrick Jaillet, and Amedeo Odoni, "Chapter 3 stochastic and dynamic networks and routing," in *Network Routing*, C.L. Monma M.O. Ball, T.L. Magnanti and G.L. Nemhauser, Eds., vol. 8 of *Handbooks in Operations Research and Management Science*, pp. 141–295. Elsevier, 1995. 162

[85] Robert B. Dial, "Autonomous dial-a-ride transit introductory overview," *Transportation Research Part C: Emerging Technologies*, vol. 3, no. 5, pp. 261–275, 1995. 162

[86] Eric Taillard, Philippe Badeau, Michel Gendreau, Francois Guertin, and Jean-Yves Potvin, "A Tabu Search Heuristic for the Vehicle Routing Problem with Soft Time Windows," *TRANSPORTATION SCIENCE*, vol. 31, no. 2, pp. 170–186, 1997. 162

[87] Timon C. Du, Eldon Y. Li, and Defrose Chou, "Dynamic vehicle routing for online b2c delivery," *Omega*, vol. 33, no. 1, pp. 33–45, February 2005. 162

[88] Marius M. Solomon, "Algorithms for the vehicle routing and scheduling problems with time window constraints," *Operations Research*, vol. 35, pp. 254–265, 1987. 162

[89] Olli Braysy and Michel Gendreau, "Vehicle Routing Problem with Time Windows, Part I: Route Construction and Local Search Algorithms," *TRANSPORTATION SCIENCE*, vol. 39, no. 1, pp. 104–118, 2005. 162

[90] Olli Braysy and Michel Gendreau, "Vehicle Routing Problem with Time Windows, Part II: Metaheuristics," *TRANSPORTATION SCIENCE*, vol. 39, no. 1, pp. 119–139, 2005. 162

[91] Dimitris Bertsimas and Garrett Van Ryzin, "Stochastic and dynamic vehicle routing in the euclidean plane with multiple capacitated vehicles," Working papers 3287-91., Massachusetts Institute of Technology (MIT), Sloan School of Management, 1991. 162

[92] M. Pavone, N. Bisnik, E. Frazzoli, and V. Isler, "A stochastic and dynamic vehicle routing problem with time windows and customer impatience," *Mobile Networks and Applications*, vol. 14, pp. 350–364, June 2009. 163

[93] Chung-Yi Lin, Sheng-Wen Shih, Yi-Ping Hung, and Gregory Y. Tang, "A new approach to automatic reconstruction of a 3-d world using active

stereo vision," *Computer Vision and Image Understanding*, vol. 85, no. 2, pp. 117–143, 2002. 163

[94] Konstantinos A. Tarabanis, Peter K. Allen, and Roger Y. Tsai, "A survey of sensor planning in computer vision," *IEEE Transactions on Robotics and Automation*, vol. 11, no. 1, pp. 86–104, February 1995. 163, 172

[95] Cregg K. Cowan and Peter D. Kovesi, "Automatic sensor placement from vision task requirements," *IEEE Transactions on Pattern Analysis and Machine Intelligence*, vol. 10, no. 3, pp. 407–416, May 1988. 163

[96] Steven Abrams, Peter K. Allen, and Konstantinos Tarabanis, "Computing camera viewpoints in an active robot work-cell," *International Journal of Robotics Research*, vol. 18, pp. 267–285, 1999. 164

[97] Takashi Matsuyama, Toshikazu Wada, and Shogo Tokai, "Active image capturing and dynamic scene visualization by cooperative distributed vision," in *Advanced Multimedia Content Processing*, Shojiro Nishio and Fumio Kishino, Eds., vol. 1554 of *Lecture Notes in Computer Science*, pp. 252–288. Springer US, 1999. 164

[98] Ardevan Bakhtari, Matthew Mackay, and Beno Benhabib, "Active-vision for the autonomous surveillance of dynamic, multi-object environments," *Journal of Intelligent and Robotic Systems*, vol. 54, pp. 567–593, 2009. 164

[99] Bryan Horling, Régis Vincent, Roger Mailler, Jiaying Shen, Raphen Becker, Kyle Rawlins, and Victor Lesser, "Distributed sensor network for real time tracking," in *Proceedings of the Fifth International Conference on Autonomous agents (AGENTS)*, New York, NY, USA, 2001, pp. 417–424, ACM. 164

[100] John R. Spletzer and Camillo J. Taylor, "Dynamic sensor planning and control for optimally tracking targets," *International Journal of Robotic Research*, vol. 22, no. 1, pp. 7–20, 2003. 164

[101] Mohamed Kamel and Lovell Hodge, "A coordination mechanism for model-based multi-sensor planning," in *Proceedings of the IEEE International Symposium on Intelligent Control*, 2002, pp. 709–714. 164

[102] Hongjun Zhou and Shigeyuki Sakane, "Sensor planning for mobile robot localization using bayesian network representation and inference," in *Proceedings of IEEE/RSJ International Conference on Intelligent Robots and Systems*, 2002, vol. 1, pp. 440–446. 164

[103] Emanuele Trucco, Manickam Umasuthan, Andrew M. Wallace, and Vito Roberto, "Model-based planning of optimal sensor placements for inspection," *IEEE Transactions on Robotics and Automation*, vol. 13, no. 2, pp. 182–194, April 1997. 165

[104] Trevor Darrell and Alex P. Pentland, "Attention-driven expression and gesture analysis in an interactive environment," in *Proceedings of the International Workshop on Automatic Face and Gesture Recognition*, 1995, pp. 135–140. 165

[105] Steven G. Goodridge, Ren C. Luo, and Michael G. Kay, "Multi-layered fuzzy behavior fusion for real-time control systems with many sensors," in *IEEE Transactions on Industrial Electronics*, 1996, vol. 43, pp. 387–394, no. 3. 165

[106] Alan LaMont Pope, *The CORBA reference guide: understanding the Common Object Request Broker Architecture*, Addison-Wesley Longman Publishing Co., Inc., Boston, MA, USA, 1998. 167

[107] Douglas C. Schmidt and Fred Kuhns, "An overview of the real-time corba specification," *Computer*, vol. 33, pp. 56–63, 2000. 167

[108] Minimum CORBA, "Minimum corba specification," http://www.omg.org/technology/documents/, [Online; accessed 14-December-2010]. 167

[109] Roger Sessions, *COM and DCOM: Microsoft's vision for distributed objects*, John Wiley & Sons, Inc., New York, NY, USA, 1998. 167

[110] Esmond Pitt and Kathy McNiff, *The CORBA reference guide: understanding the Common Object Request Broker Architecture*, Addison-Wesley Longman Publishing Co., Inc., Boston, MA, USA, 2001. 167

[111] Yang Yu, Bhaskar Krishnamachari, and V.K. Prasanna, "Issues in designing middleware for wireless sensor networks," *IEEE Network*, vol. 18, no. 1, pp. 15–21, January 2004. 167

[112] Mohammad M. Molla and Sheikh Iqbal Ahamed, "A survey of middleware for sensor network and challenges," in *Proceedings of the International Conference Workshops on Parallel Processing*, Washington, DC, USA, 2006, pp. 223–228, IEEE Computer Society. 168

[113] Philip Levis, Sam Madden, Joseph Polastre, Robert Szewczyk, Alec Woo, David Gay, Jason Hill, Matt Welsh, Eric Brewer, and David Culler, "Tinyos: An operating system for sensor networks," in *Ambient Intelligence*, Werner Weber, Jan M. Rabaey, and Emile Aarts, Eds. Springer Verlag, 2004. 168

[114] Christian Becker, Gregor Schiele, Holger Gubbels, and Kurt Rothermel, "BASE - a micro-broker-based middleware for pervasive computing," in *Proceedings of the First IEEE International Conference on Pervasive Computing and Communications (PerCom)*, March 2003, pp. 443–451. 168

[115] Florian Mösch, Marek Litza, Adam El Sayed Auf, Erik Maehle, Karl-Erwin Großpietsch, and Werner Brockmann, "Orca - towards an organic robotic control architecture," in *IWSOS/EuroNGI*, 2006, pp. 251–253. 168

[116] Wolfgang Trumler, Faruk Bagci, Jan Petzold, and Theo Ungerer, "AMUN-autonomic middleware for ubiquitous environments applied to the smart doorplate project," *Advanced Engineering Informatics*, vol. 19, pp. 243–252, July 2005. 168

[117] Andreas Pietzowski, Wolfgang Trumler, and Theo Ungerer, "An artificial immune system and its integration into an organic middleware for self-

protection," in *Proceedings of the 8th annual Conference on Genetic and Evolutionary Computation (GECCO)*, New York, NY, USA, 2006, pp. 129–130, ACM. 168

[118] Martin Hoffmann, *System Management Algorithms for Distributed Vision Networks*, Ph.D. thesis, Institute of Systems Engineering / SRA, University of Hannover, 2010. 168

[119] Hamid Aghajan and Andrea Cavallaro, *Multi-Camera Networks: Principles and Applications*, Academic Press, 2009. 169

[120] Joseph O'Rourke, *Art gallery theorems and algorithms*, Oxford University Press, Inc., New York, NY, USA, 1987. 170

[121] V. Chvatal, "A combinatorial theorem in plane geometry," *Journal of Combinatorial Theory Series B*, vol. 18, pp. 39–41, 1975. 170

[122] D. T. Lee and Arthur K. Lin, "Computational complexity of art gallery problems," *IEEE Transactions on Information Theory*, vol. 32, no. 2, pp. 276–282, 1986. 170

[123] Héctor González-banos, "A randomized art-gallery algorithm for sensor placement," in *Proceedings of the Seventeenth annual Symposium on Computational Geometry (SCG)*, 2001, pp. 232–240. 170

[124] Danny Yang, Jaewon Shin, Ali Ozer Ercan, and Leonidas Guibas, "Sensor tasking for occupancy reasoning in a camera network," in *Proceedings of IEEE/ICST 1st Workshop on Broadband Advanced Sensor Networks (BASENETS)*, 2004. 170

[125] Pere-Pau Vázquez, Miquel Feixas, Mateu Sbert, and Wolfgang Heidrich, "Viewpoint selection using viewpoint entropy," in *Proceedings of the Vision Modeling and Visualization Conference (VMV)*. 2001, pp. 273–280, Aka GmbH. 170

[126] Jeff Williams and Won-Sook Lee, "Interactive virtual simulation for multiple camera placement," in *IEEE International Workshop on Haptic Audio Visual Environments and Their Applications*, 2006. 170

[127] Anurag Mittal and Larry S. Davis, "A general method for sensor planning in multi-sensor systems: Extension to random occlusion," *International Journal of Computer Vision*, vol. 76, pp. 31–52, January 2008. 170

[128] Robert Bodor, Andrew Drenner, Paul Schrater, and Nikolaos Papanikolopoulos, "Optimal camera placement for automated surveillance tasks," *Journal of Intelligent and Robotic Systems*, vol. 50, pp. 257–295, 2007. 170

[129] Ali O. Ercan, Danny B. Yang, Abbas El Gamal, and Leonidas J. Guibas, "Optimal placement and selection of camera network nodes for target localization," in *DCOSS*, 2006, pp. 389–404. 170

[130] Enrique Dunn and Gustavo Olague, "Pareto optimal camera placement for automated visual inspection," in *Proceedings of IEEE/RSJ International Conference on Intelligent Robots and Systems (IROS)*, August 2005, pp. 3821–3826. 170

[131] E. Hörster and R. Lienhart, "On the optimal placement of multiple visual sensors," in *Proceedings of the 4th ACM international workshop on Video Surveillance and Sensor Networks (VSSN)*, New York, NY, USA, 2006, pp. 111–120, ACM. 170

[132] Mohammad Al Hasan, Krishna Ramachandran, and John Mitchell, "Optimal placement of stereo sensors," *Optimization Letters*, vol. 2, pp. 99–111, 2008. 170

[133] Ugur Murat Erdem and Stan Sclaroff, "Optimal placement of cameras in floorplans to satisfy task requirements and cost constraints," in *Proceedings of OMNIVIS Workshop*, 2004. 170

[134] Tao Zhao and Ram Nevatia, "Tracking multiple humans in complex situations," *IEEE Transactions on Pattern Analysis and Machine Intelligence*, vol. 26, pp. 1208–1221, 2004. 171

[135] N. Krahnstoever, P. Tu, T. Sebastian, A. Perera, and R. Collins, "Multi-view detection and tracking of travelers and luggage in mass transit environments," in *Proceedings of the 9th IEEE International Workshop on Performance Evaluation of Tracking and Surveillance and CVPR*, June 2006. 171, 173

[136] Peter Tu, Fred Wheeler, Nils Krahnstoever, Thomas Sebastian, Jens Rittscher, Xiaoming Liu, and Amitha Perera, "Surveillance video analytics for large camera networks," *SPIE Newsletter*, 2007. 171

[137] Ting Yu, Ying Wu, N.O. Krahnstoever, and P.H. Tu, "Distributed data association and filtering for multiple target tracking," in *Proceedings of IEEE Conference on Computer Vision and Pattern Recognition (CVPR)*, June 2008, pp. 1–8. 171

[138] John Krumm, Steve Harris, Brian Meyers, Barry Brumitt, Michael Hale, and Steve Shafer, "Multi-camera multi-person tracking for easyliving," in *Proceedings of the Third IEEE International Workshop on Visual Surveillance (VS)*, Washington, DC, USA, 2000, p. 3, IEEE Computer Society. 171

[139] Martin Hoffmann, Michael Wittke, Yvonne Bernard, Ramin Soleymani, and Jörg Hähner, "DMCtrac: Distributed multi camera tracking," in *Proceedings of Second ACM/IEEE International Conference on Distributed Smart Cameras (ICDSC)*, September 2008, pp. 1–10. 171

[140] Martin Hoffmann, Michael Wittke, Jörg Hähner, and Christian Müller-Schloer, "Spatial partitioning in self-organising camera systems," *IEEE Journal of Selected Topics in Signal Processing*, vol. 2, pp. 480–492, August 2008. 171

[141] Shaogang Gong, Jamie Sherrah, and Jeffrey Ng, "On the semantics of visual behavior, structured events and trajectories of human action," *Image and Vision Computing*, vol. 20(12), pp. 873–888, 2002. 171

[142] Gerard Medioni, Isaac Cohen, and Ram Nevatia, "Event detection and analysis from video streams," *IEEE Transactions on Pattern Analysis and Machine Intelligence*, vol. 23, no. 8, pp. 873–889, August 2001. 171

[143] Andrew Senior, Arun Hampapur, and M. Lu, "Acquiring multi-scale images by pan-tilt-zoom control and automatic multi-camera calibration," in *Proceedings of Seventh IEEE Workshops on Application of Computer Vision (WACV/MOTIONS)*, January 2005, vol. 1, pp. 433–438. 171

[144] S. J. D. Prince, J. H. Elder, Y. Hou, and M. Sizinstev, "Pre-attentive face detection for foveated wide-field surveillance," in *Proceedings of Seventh IEEE Workshops on Application of Computer Vision (WACV/MOTIONS)*, January 2005, vol. 1, pp. 439–446. 171

[145] Siva Ram, K. R. Ramakrishnan, P. K. Atrey, V. K. Singh, and M. S. Kankanhalli, "A design methodology for selection and placement of sensors in multimedia surveillance systems," in *Proceedings of the 4th ACM international workshop on Video Surveillance and Sensor Networks (VSSN)*, New York, NY, USA, 2006, pp. 121–130, ACM. 171

[146] L. Marchesotti, L. Marcenaro, and C. Regazzoni, "Dual camera system for face detection in unconstrained environments," in *Proceedings of the International Conference on Image Processing (ICIP)*, September 2003, vol. 1, pp. 681–684. 171

[147] Xuhui Zhou, Robert Collins, Takeo Kanade, and Peter Metes, "A master-slave system to acquire biometric imagery of humans at distance," in *ACM International Workshop on Video Surveillance*, ACM, Ed., November 2003. 172

[148] Julie Badri, Christophe Tilmant, Jean-Marc Lavest, Quonc-Cong Pham, and Patrick Sayd, "Camera-to-camera mapping for hybrid pan-tilt-zoom sensors calibration," in *Proceedings of the 15th Scandinavian Conference on Image Analysis (SCIA)*, Berlin, Heidelberg, 2007, pp. 132–141, Springer-Verlag. 172

[149] P. Peixoto, J. Batist, and H. Aralujo, "Real-time active visual surveillance by integrating peripheral motion detection with foveated tracking," *IEEE Workshop on Visual Surveillance*, vol. 0, pp. 18, 1998. 172

[150] Ser-Nam Lim, L. S. Davis, and A. Elgammal, "Scalable image-based multi-camera visual surveillance system," in *Proceedings of the International IEEE Conference on Advanced Video and Signal Based Surveillance*, July 2003, pp. 205–212. 172

[151] Arun Hampapur, Sharat Pankanti, Andrew Senior, Ying-Li Tian, Lisa Brown, and Ruud Bolle, "Face cataloger: multi-scale imaging for relating identity to location," in *Proceedings of the International IEEE Conference on Advanced Video and Signal Based Surveillance*, July 2003, pp. 13–20. 172

[152] Faisal Z. Qureshi and Demetri Terzopoulos, "Surveillance camera scheduling: a virtual vision approach," in *Proceedings of the third ACM international workshop on Video Surveillance and Sensor Networks (VSSN)*, New York, NY, USA, 2005, pp. 131–140, ACM. 172

[153] Alberto Del Bimbo and Federico Pernici, "Towards on-line saccade planning for high-resolution image sensing," *Pattern Recognition Letters*, vol. 27, no. 15, pp. 1826–1834, 2006, Vision for Crime Detection and Prevention. 172

[154] Yiming Li and B. Bhanu, "Utility-based dynamic camera assignment and hand-off in a video network," in *Proceedings of the Second ACM/IEEE International Conference on Distributed Smart Cameras (ICDSC)*, September 2008, pp. 1–9. 172

[155] D.J. Cook, P. Gmytrasiewicz, and L.B. Holder, "Decision-theoretic cooperative sensor planning," *IEEE Transactions on Pattern Analysis and Machine Intelligence*, vol. 18, no. 10, pp. 1013–1023, October 1996. 173

[156] N. Ukita and T. Matsuyama, "Real-time cooperative multi-target tracking by communicating active vision agents," in *Proceedings of the 16th*

International Conference on Pattern Recognition, 2002, vol. 2, pp. 14–19. 173

[157] Zhen Jia, A. Balasuriya, and S. Challa, "Recent developments in vision based target tracking for autonomous vehicles navigation," in *Proceeedings of the International IEEE Conference on Intelligent Transportation Systems (ITSC)*, September 2006, pp. 765–770. 173

[158] Jr. Mazo, M., A. Speranzon, K.H. Johansson, and Xiaoming Hu, "Multi-robot tracking of a moving object using directional sensors," in *Proceedings of the IEEE International Conference on Robotics and Automation (ICRA)*, April 2004, vol. 2, pp. 1103–1108. 174

[159] A. Betser, P. Vela, and A. Tannenbaum, "Automatic tracking of flying vehicles using geodesic snakes and kalman filtering," in *Proceedings of the 43rd IEEE Conference on Decision and Control (CDC)*, December 2004, vol. 2, pp. 1649–1654. 174

[160] Wee Kiat Ng, G.S.B. Leng, and Yee Leong Low, "Coordinated movement of multiple robots for searching a cluttered environment," in *Proceedings of IEEE/RSJ International Conference on Intelligent Robots and Systems (IROS)*, September 2004, vol. 1, pp. 400–405. 174

[161] Jana van Greunen and Jan Rabaey, "Lightweight time synchronization for sensor networks," in *Proceedings of the 2nd ACM International Conference on Wireless Sensor Networks and Applications (WSNA)*, New York, NY, USA, 2003, pp. 11–19, ACM. 175

[162] Jérémie Allard, Valérie Gouranton, Guy Lamarque, Emmanuel Melin, and Bruno Raffin, "Softgenlock: active stereo and genlock for pc cluster," in *Proceedings of the workshop on Virtual Environments (EGVE)*, New York, NY, USA, 2003, pp. 255–260, ACM. 176

[163] Kay Römer, "Time synchronization in ad hoc networks," in *Proceedings of the 2nd ACM international symposium on Mobile ad hoc networking and computing (MobiHoc)*, New York, NY, USA, 2001, pp. 173–182, ACM. 177

[164] S. Ping, "Delay Measurement Time Synchronization for Wireless Sensor Networks," Intel Research IRB-TR-03-0133, Networked and Embedded Systems Lab (NESL), University of California, Los Angeles (UCLA), 2003. 177

[165] Qun Li and Daniela Rus, "Global clock synchronization in sensor networks," *IEEE Transactions on Computers*, vol. 55, pp. 214–226, February 2006. 177

[166] J. Yan and M. Pollefeys, "Video synchronization via space-time interest point distribution," *Advanced Concepts for Intelligent Vision Systems*, 2004. 178

[167] S. Kuthirummal, C.V. Jawahar, and P.J. Narayanan, "Video frame alignment in multiple views," in *Proceedings of the International Conference on Image Processing*, 2002, vol. 3, pp. 357–360. 178

[168] L. Lee, R. Romano, and G. Stein, "Monitoring activities from multiple video streams: establishing a common coordinate frame," *Transactions on Pattern Analysis and Machine Intelligence*, vol. 22, no. 8, pp. 758–767, 2000. 178

[169] Senem Velipasalar and Wayne Wolf, "Frame-level temporal calibration of video sequences from unsynchronized cameras by using projective invariants," in *Proceedings of the International IEEE Conference on Advanced Video and Signal Based Surveillance (AVSS)*, September 2005, pp. 462–467. 179

[170] Jeremy Elson, Lewis Girod, and Deborah Estrin, "Fine-grained network time synchronization using reference broadcasts," in *Proceedings of the 5th symposium on Operating systems design and implementation of ACM SIGOPS Operating Systems Review (OSDI)*, New York, NY, USA, 2002, vol. 36, pp. 147–163, ACM. 179

[171] Santashil Palchaudhuri, Amit Saha, and David B. Johnson, "Probabilistic Clock Synchronization Service in Sensor Networks," Tech. Rep., Department of Computer Science, 2003. 179

[172] Michael Mock, Reiner Frings, Edgar Nett, and Spiro Trikaliotis, "Continuous Clock Synchronization in Wireless Real-Time Applications," in *Proceedings of the 19th IEEE Symposium on Reliable Distributed Systems (SRDS)*, Washington, DC, USA, 2000, p. 125, IEEE Computer Society. 179

[173] Weilian Su and Ian F. Akyildiz, "Time-diffusion synchronization protocol for wireless sensor networks," *IEEE/ACM Transactions on Networking (TON)*, vol. 13, no. 2, pp. 384–397, 2005. 180

[174] David W. Casbeer, Derek B. Kingston, Al W. Beard, Timothy W. Mclain, Sai ming Li, and Raman Mehra, "Cooperative forest fire surveillance using a team of small unmanned air vehicles," *International Journal of Systems Sciences*, vol. 37, pp. 360, 2006. 180

i want morebooks!

Buy your books fast and straightforward online - at one of world's fastest growing online book stores! Environmentally sound due to Print-on-Demand technologies.

Buy your books online at
www.get-morebooks.com

Kaufen Sie Ihre Bücher schnell und unkompliziert online – auf einer der am schnellsten wachsenden Buchhandelsplattformen weltweit! Dank Print-On-Demand umwelt- und ressourcenschonend produziert.

Bücher schneller online kaufen
www.morebooks.de

 VDM Verlagsservicegesellschaft mbH
Heinrich-Böcking-Str. 6-8   Telefon: +49 681 3720 174   info@vdm-vsg.de
D - 66121 Saarbrücken   Telefax: +49 681 3720 1749   www.vdm-vsg.de

Printed by Books on Demand GmbH, Norderstedt / Germany